W9-BNP-714

IMAGINING NUMBERS

IMAGINING NUMBERS

(particularly the square root of minus fifteen)

PICADOR
FARRAR, STRAUS AND GIROUX
NEW YORK

 For information, address Picador, 175 Fifth Avenue, New York, N.Y. 10010.

www.picadorusa.com

For information on Picador Reading Group Guides, as well as ordering, please contact the Trade Marketing department at St. Martin's Press.
Phone: 1-800-221-7945 extension 763
Fax: 212-677-7456
E-mail: trademarketing@stmartins.com

Grateful acknowledgment is made for the use of the following illustrations: *p. 140:* Vladimir Nabokov's illustration from Kafka's *The Metamorphosis,* by arrangement with the Estate of Vladimir Nabokov. All rights reserved. *p. 222:* Facsimile reproduction of Bombelli manuscript from *Per la storia e la filosofia delle matematiche* by E. Bortolotti, courtesy of Zanichelli Editore S.P.A., Bologna, Italy. Owing to limitations of space, all other acknowledgments for permission to reprint previously published material can be found on pages 269–270.

Library of Congress Cataloging-in-Publication Data

Mazur, Barry.
Imagining numbers : (particularly the square root of minus fifteen) / Barry Mazur.
p. cm.
Includes bibliographical references and index.
ISBN 0-312-42187-7
1 Numbers, Complex. I. Title.

QA255 .M39 2002
512—dc21 2002075402

First published in the United States by Farrar, Straus and Giroux

First Picador Edition: February 2004

10 9 8 7 6 5 4 3 2 1

To Gretchen

CONTENTS

PART III

PREFACE

1, 2, 3, . . . The "counting numbers" are part of us. We know them forward and backward. Babies as young as five months old, psychologists claim,[1] are sensitive to the difference between 1 + 1 and 2 − 1. We sing numbers, counting up the days of Christmas and counting down to the poignant monotheism of "One is one and all alone and evermore shall be so."

Our ancestors have added to this repertoire and reckoned with zero and the negative numbers, which were sometimes referred to as fictions (*fictae*) before they gained familiarity.

All these together constitute what we call the *whole numbers*,

$$\ldots, -2, -1, 0, +1, +2, \ldots$$

More formally, they are called *integers*, from the Latin adjective meaning "whole, untouched, unharmed."

"Whole," "untouched"; their very name hints that integers *can be* touched, or fractionated. Indeed they can be, and when they are, we get the larger array of numbers that are *fractions*, ratios of whole numbers.

Fractions, as their notation vividly displays, also stand for proportions (think of $\frac{1}{2} = \frac{2}{4}$ as "one is to two as two is to four") and for actions (think of $\frac{1}{2}$ as "halving," ready to cut in half anything that follows it).

To bring fractions into line, we express them as decimals ($\frac{1}{2}$ = 0.5000000 . . .). The modern world gives us much experience with decimals to a high degree of accuracy—to "many decimal places"; mathematics, as always, goes all the way, happy to deal with numbers with complete accuracy—to *infinitely* many decimal places. Numbers represented by infinitely many decimal places, whether they are fractions or not, are called *real numbers.*

But the telltale adjective *real* suggests two things: that these numbers are somehow real to us and that, in contrast, there are *unreal* numbers in the offing. These are the *imaginary numbers.*

The *imaginary* numbers are well named, for there is some imaginative work to do to make them as much a part of us as the real numbers we use all the time to measure for bookshelves.

This book began as a letter to my friend Michel Chaouli. The two of us had been musing about whether or not one could "feel" the workings of the imagination in its various labors.[2] Michel had also mentioned that he wanted to "imagine imaginary numbers." That very (rainy) evening, I tried to work up an explanation of the idea of these numbers, still in the mood of our conversation.

The text of my letter was the welcome excuse for continued conversation with a number of friends, many of whom were humanists interested in understanding what it means to imagine the square root of negative numbers. The successive revisions and expansions of my letter were shaped by their questions, comments, critiques, and insights. This book, then, is written for people who have no training in mathematics and who may not have actively thought about mathematics since high school, or even during it, but who may wish to experience an act of mathematical imagining and to consider how such an experience compares with the imaginative work involved in reading and understanding a phrase in a poem. Of course, poetry and mathematics are far apart. All the more glorious, then, for people at home in the imaginative life of poetry to use their insight and sensibility to witness the imagination at work in mathematics.

Although no particular mathematical knowledge is necessary, pencil and paper are good to have at hand, to make a few calculations (multiplying small numbers, mostly). The operation of multiplication itself is something we will be looking at. There are three standard ways of denoting the act of multiplication: by a *cross* ×, by a *centered dot* ·, or, when there is no ambiguity, by *simple juxtaposition* of the objects to be multiplied. Which notation we use reflects where we wish to direct our attention: the equation

$$2 \times 3 = 6$$

emphasizes the *act* of multiplying 2 times 3, whereas

$$2 \cdot 3 = 6$$

focuses on the *result*, 6, of that operation. Nevertheless, despite this difference in nuance, both equations, $2 \times 3 = 6$ and $2 \cdot 3 = 6$, are saying the same thing. When we deal with an unknown quantity X, here are three equivalent ways of denoting 5 times that unknown quantity:

$$5 \times X = 5 \cdot X = 5X.$$

Again, we write $5 \times X$ if we want to emphasize the act of multiplying and $5 \cdot X$ or $5X$ if we want to emphasize the result; and that last variant notation, juxtaposition, is used for visual conciseness.

This book has footnotes and endnotes. Some of the endnotes are side comments requiring more mathematical background than is assumed in the text.

PART I

1

THE IMAGINATION AND SQUARE ROOTS

1. Picture this.

Picture Rodin's *Thinker*, crouched in mental effort. He has his supporting right elbow propped *not* on his right thigh, as you or I would have placed our right elbow, but rather on his left thigh,[1] which bolts him into an awkward striving, his muscles tense with thought. But does he, can we, really *feel* our imaginative faculty at work, striving toward, and then finally achieving, an act of the imagination?

Consider the range of our imaginative experiences. Consider, for example, how immediate is the experience of imagining what we read. Elaine Scarry has remarked that there is *no* "felt experience" corresponding to this imaginative act.[2] We experience, of course, the *effect* of what we are reading. Scarry claims that if we read a phrase like the yellow of the tulip[3]

we form, perhaps, the image of it in our mind's eye and experience whatever emotional effect that image produces within us. But, says Scarry, we have no *felt experience* of coming to form that image. We will return to this idea later.

Perhaps one should contrast reading with trying to think something up for ourselves. Rainer Maria Rilke's comment on the working of our imagination,

We are the bees of the invisible[4]

paints our imaginative quests as not entirely *unfelt experiences* (following Scarry), but not *contortions* (following Rodin) either. Our gathering of the honey of the imaginative world is not immediate; it takes work. But though it requires traveling some distance, merging with something not of our species, communicating by dance to our fellow creatures what we've done and where we've been, and, finally, bringing back that single glistening drop, it is an activity we do without contortion. It is who we bees are.

Thinking about the *imagination imagining* is made difficult by the general swiftness and efficacy of that faculty. The imagination is a fleet genie at your service. You want an elephant? Why, there it is:

(picture your elephant here)

You read "the yellow of the tulip."

And, again, there it is: a calligraphed swath of yellow on your mental movie screen.

More telling, though, are the other moments of thought, when our genie is not so surefooted. Moments composed half of bewilderment and half of expectation; moments, for example, when some new image, or viewpoint, is about to reveal itself to us. But it resists emerging. We are forced to angle for it.

At those times, it is as if the waters of the imagination are roiling; you have cast your fishing line from a somewhat shaky boat, and you feel a tug on that line, but have no clear sense what you have hooked onto. Bluefish, old boot, or some underwater species never before seen? But you definitely feel the tug.

I want to think about the inner articulations of our imaginative life by "re"-experiencing a particular example of such a tug. The example I propose to consider occurs in the history of mathematics. It might be described as a *moment of restless anticipation in the face of a slowly emerging act of imagining. Moment*, though, is not the right word here, for the period, rather, stretches over three centuries. And *anticipation* carries too progressivist and perhaps too personal a tone, for this "act" doesn't take place fully in any single mind. There are many "bees of the invisible" in the original story.

If we are successful, we will be reenacting, for ourselves, the imagining of a concept that, for the original thinkers, had never been seen or thought before, and that seemed to lie athwart things seen or thought before.* Of course, thinking about things never thought before is the daily activity of thought, certainly in art or science. The cellist Yo-Yo Ma has suggested that the job of the artist is to go to the edge and report back.[5] Here is how Rilke expressed a similar sentiment: "Works of art are indeed always products of having-been-in-danger, or having-gone-to-the-very-end in an experience, to where one can go no further."[6]

In contrast to the instantly imaginable "yellow of the tulip," the *square root of negative quantities* was a concept in common use in mathematics for over three hundred years before a satisfactory geometric under-

*This sentence echoes the caption of an old cartoon in which a child is pursued by a demon of his imagination and cries out, "It looks very much like something I have never seen before!"

standing of it was discovered. If you deal *exclusively with positive quantities*, you have less of a challenge in coming to grips with square roots: the *square root of a positive number* is just a quantity whose square is that number.

Any positive number has only one (positive) square root. The square root of 4, for example, is 2. What is the square root of 2? We know, at the very least, that its square is 2. Using the equation that asserts this,

$$(\sqrt{2})^2 = \sqrt{2} \cdot \sqrt{2} = 2,$$

try your hand at estimating $\sqrt{2}$. Is it smaller than $\frac{3}{2}$? Do you see why $\sqrt{3} \cdot \sqrt{5} = \sqrt{15}$?

Square roots are often encountered geometrically, as lengths of lines. We will see shortly, for example, that $\sqrt{2}$ is the length of the diagonal of a square whose sides have length 1.

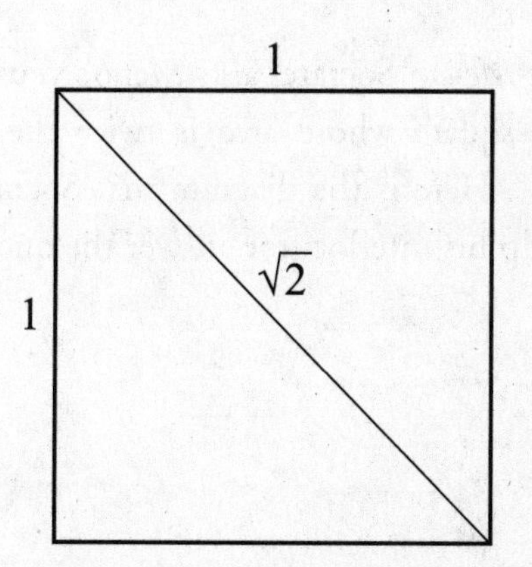

Also, if we have a square figure whose area is known to be A square feet, then the length of each of its sides, as in the diagram below, is $\sqrt{A}$ feet.

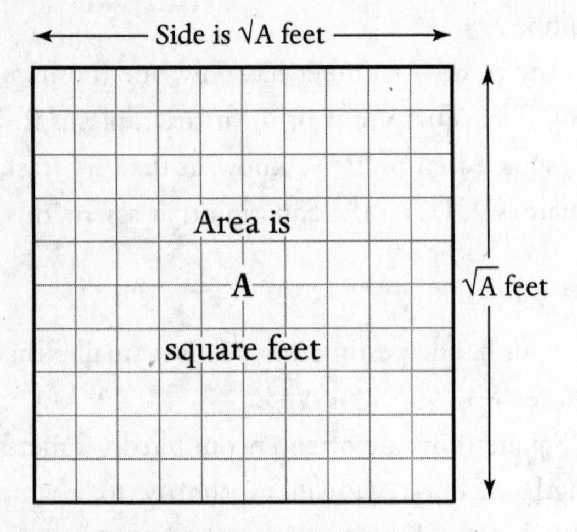

The square root as "side"

Suppose that each box in this diagram has an area equal to 1 square foot. There are a hundred boxes, so $A = 100$, and the dimensions of the large square are $\sqrt{A}$ by $\sqrt{A}$—that is, 10 by 10.

In Plato's *Meno*,[7] Socrates asks Meno's young slave to construct a square whose area is twice the area of a given square. Here is the diagram that Socrates finally draws to help his interlocutor answer the question:

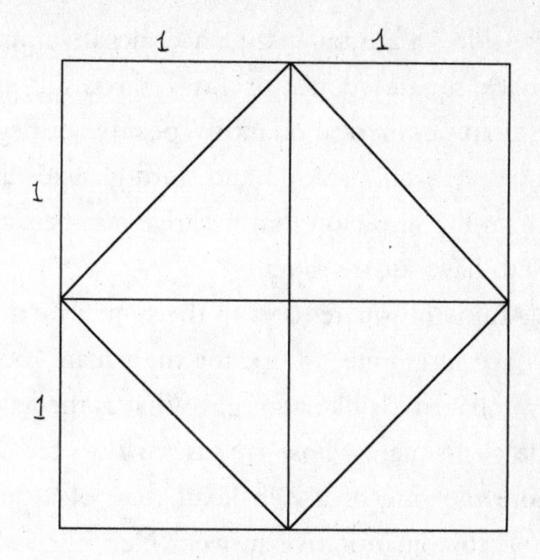

The profile of this diagram is a 2×2 square (whose area is therefore 4) built out of four 1×1 squares (each of area 1). But in its midst, we can pick out a catercorner square (standing, as it seems, on one of its corners). By rearranging the triangular pieces that make up the diagram, can you see, as Socrates' young friend in the *Meno* did, that the catercorner square has area 2, and therefore each of its sides has length $\sqrt{2}$?

The sides of the catercorner square play a double role: they are also the diagonals of the small (1×1) squares. So, as promised a few paragraphs earlier, we see $\sqrt{2}$ as the length of the diagonal drawn in a square whose sides are of length 1.

The early mathematicians thought of the square root as a "side"; the sixteenth-century Italians would at times simply refer to the square root of a number as its *lato*, its "side." Thus, at first glance, negative numbers don't have square roots, for (as I discuss later) the square of any numerical quantity (positive or negative) is positive. In fact, a second and third glance will tend to confirm the suspicion that negative numbers are not *entitled* to have square roots.

If we think of square roots in the geometric manner, as we have just done, to ask for the square root of a negative quantity is like asking: "What is the length of the side of a square whose area is *less than zero*?" This has more the ring of a Zen koan than of a question amenable to a quantitative answer.[8] Nevertheless, these seemingly nonexistent square roots were, early on, seen to be *useful*. But the first users of square roots of negative numbers were queasy about the practice of invoking such airy objects. These strange square roots were called *imaginary numbers*, meaning they were difficult to place among *real* mathematical objects.

And then, an astonishingly satisfying image of these square roots emerged. A way of *imagining* these otherwise unpicturable "numbers" was found independently, and almost simultaneously, by two, or possibly three (or more), people.* What a dramatic act: to find a

*A friend suggested that since none of those directly involved in the publication of this discovery had any other significant mathematical contribution to their credit (with the exception of Legendre, who, as we shall see, plays a curious role)—that is, since all these individuals are peripheral to the intense mathematical progress of the time (the end of the eighteenth century)—it is possible that the "pictorial image" they came

home in our imagination for such an otherwise troublesome concept!

This "way of imagining" has become our common intellectual property. It and the numbers it helped us imagine have found thorough and ubiquitous use, not only by mathematicians but by every engineer who works with the calculus, by every physicist.

The aim of this book is not to give a historical account.* Rather, it is to re-create, in ourselves, the shift of mathematical thought that makes it possible to imagine these numbers.

Poetry, to be sure, has "shifts of thought" at its core, the "turn" of the poem, in both its viewpoint and its typography, being celebrated in the word *verse*. Poetry demands our paying attention to these turns. For people who pay such attention while reading poetry but who have never done anything similar with mathematics, I hope the style of presentation I have adopted—which passes back and forth between reflections on the imaginative work of thinking about poetry and thinking about mathematics—will be helpful.

In proceeding with our mathematical theme, we

---

up with was, in fact, "in the air," was in the "public domain," at least to the extent that the "public" included Euler and his colleagues. In any event, plucking such coins out of the air is a pretty good trick, which, with luck, we too will do in subsequent pages.

*See the annotated bibliography at the end of the book for a list of sources that provide a systematic logical or historical account of the concept of number.

want first to feel the uncomfortableness of the early mathematicians who encounter imaginary numbers; then to sense the possibility that some shift, some new viewpoint in thinking about number, may help to tame the concept of imaginary number; then to be conscious of the emergence of this viewpoint within ourselves. Finally, we will see that our new attitude toward number unifies otherwise disparate intuitions and helps us interpret an amazing formula that perplexed sixteenth-century mathematicians.

As for prerequisites, the less mathematics you know, the better prepared you are for the task ahead. To follow the mathematics presented here, you will only need to have the skill to perform certain simple multiplications and substitutions when the text requests this, and to allow with equanimity the occasional appearance of simple algebraic equations of the type encountered in the first weeks of high school algebra.* If you can do, or follow, the sample exercises in this endnote,[9] you are ready for the math in this book.

Let us start by considering that imaginative construct, the faculty of imagination itself.

*A comment by one of the readers of an early draft of this book led me to revise it substantially. My manuscript, the reader said, reminded him of the time he paged through the *Kama Sutra*. The *Kama Sutra* promised that a wonderful world was his if only he had (but he hadn't) sufficient flexibility and skill. The present version of this book requires neither.

2. Imagination.

A certain mathematical article opens with the invitation:

> Imagine . . . an infinite completely symmetrical array of points.[10]

In the prologue to Shakespeare's *Henry V*, the Chorus asks that you, the audience, let the actors,

> *. . . ciphers to this great accompt,*
> *On your imaginary forces work.*

Paul Scott's *The Raj Quartet* begins with a request of the reader:

> IMAGINE, then, a flat landscape, dark for the moment, but even so conveying to a girl running in the still deeper shadow cast by the wall of the Bibighar Gardens an idea of immensity.[11]

What a problematic instruction: to be told to *imagine*! What are we doing, and do we have the language to say what we are doing, when we fulfill that instruction?

Our English word *imagination* has a direct antecedent in Latin, but the earlier Latin word, which connoted "object of the imagination" (at least as a side meaning), is *visio*, whose standard meaning is "sight." For a discussion about this (and for a comprehensive history and commentary on what has been said about the imagination), see Eva Brann's majestic *The World of the Imagination—Sum and Substance.*[12] Here is Quintilian explaining the Greek origin of the Latin term *visio*:

> What the Greeks call "phantasies" we rightly term "sights" through which the images of absent things are so represented in the mind that we seem to discern and have them present.[13]

Quintilian's definition of *sights* as meaning "objects of the imagination" is a serviceable definition, as far as it goes. It includes things we have seen before but which happen to be absent. Its reach, however, does not encompass the unicorns and sphynxes that tinkers and joiners of the imagination have thrown together for us.

One might try to extend Quintilian's definition, following the lead of Jeremy Bentham, by claiming that the imagination is a faculty by which "a number of abstracted ideas are compounded into one image."[14] Bentham's definition goes a bit further than Quintilian's, but not much, for surely there are objects of thought that cannot be parsed in terms of the algebra of simple, previously known images.

And Bentham's definition, which has the imaginative faculty playing the menial role of editing table for videotapes of the mind's eye, would hardly satisfy

Wordsworth, who would prefer a loftier function of the imagination: the function of connecting mere fact with "that infinity without which there is no poetry."[15] For Wordsworth, the imaginative faculty is the transcendental alchemist that turns, for example, the "mere" gold band of a wedding ring into a symbol of eternal unity.

Quintilian, Bentham, Wordsworth, et al., notwithstanding, there are those who simply shrug off "imagination" as an " 'onomatoid,' that is, a namelike word which in fact designates nothing because it signifies too broadly."[16] Is it *one* thing, deserving of the pronoun *it*? Coleridge makes a distinction in *Biographia Literaria* between what he calls the *imagination* and its less daring sibling *fancy*, which "is indeed no other than a mode of memory emancipated from the order of time and space."[17] In some circles, the concept of the imaginative faculty (or, at least, the idea that you can say anything about it) raises philosophical suspicion; in other circles, its very mention raises religious fears. For example, a recent review of high school history textbooks reports that, to satisfy the religious right, the word *imagine* is largely banished from textbooks. An editor at McGraw-Hill is quoted as saying, "We were told to try to avoid using the word 'imagine' because people in Texas felt it was too close to the word 'magic' and therefore might be considered anti-Christian."[18]

Nevertheless, there are certain experiences of the intellect that cannot be discussed at all without grappling with the issue of the imagination.

3. Imagining what we read.

When we look at a page of writing, our mind's eye sees something quite different than the white page, the black ink. John Ashbery, in his prose poem "Whatever It Is, Wherever You Are," writes of reading:

> [T]he yellow of the tulip, for instance—will flash for a moment in such a way that after it has been withdrawn we can be sure there was no imagining, no auto-suggestion here, but at the same time it becomes as useless as all subtracted memories.

He muses about the *inventors of writing*:

> To what purpose did they cross-hatch so effectively, so that the luminous surface that was underneath is transformed into another, also luminous but so shifting and so alive with suggestiveness that it is like quicksand, to take a step there would be to fall through the fragile net of uncertainties into the bog of certainty . . .

and suggests that the images conjured by reading flash onto our mental screen and convey "certainty without heat or light."[19] For Scarry, the "vivacity" of the yellow flash of the tulip compels conviction, and the suddenness of its appearance in our mind precludes our having any "felt experience of image-making." She says that

> the imagination consists exclusively of its objects, that it is only knowable through its objects, that it is remarkable among intentional states for not being easily separable into the double structure of state and object.[20]

To give a sense of what she means by the double structure of "state and object," Scarry offers the comparison of the acts of "imagining a flower" and "fearing an earthquake." She points out that "fear of an earthquake" has two parts to it: the contemplated object of the fear, and the inner experiencing of this fear. In contrast, suggests Scarry, "imagining a flower" has as object the imagined flower, but comes along with no further, separable, inner experience of the exercise of the faculty of the imagination.

This may not be surprising, for even with the (happily) true and not imagined sensory perception of *the smell of coffee in the morning*, one has the inner experience of the smell of that coffee without any palpable separate experience of the exercise of one's sense of smell.

Scarry herself, I should emphasize, does not alight for too long upon these ideas. For those who have not read her prose-poem essay, I wouldn't want to ruin its dramatic momentum by revealing how it evolves from this theme, or to reveal its further surprises. But can we, as a modest test of Scarry's claim, catch glimpses of our *imagination at work*? With this in mind, let us turn to the initial setting of our mathematical story.

4. Mathematical problems and square roots.

As already hinted, square roots show up as answers to even some of the simplest of geometric questions. And if your appetite for mathematical problems grows, you find, as did the sixteenth-century Italian algebraists, more complicated numerical quantities like

$$\sqrt{\sqrt{52}+2}$$

(this one happens to be roughly 3.03) appearing routinely as *solutions.*[21]* Reading these Italian mathematicians, you can only have admiration for the tongue-twisting lengths to which they went to indulge their tastes for mathematical puzzles, which were often allowed to masquerade as practical(?!) problems:

> A certain king sent 128,000 *aurei* to the proconsul who was leading his army so that he might hire 7000

*The general notation for square root, cube root, fourth root, and so forth, is $\sqrt[2]{\ }$, $\sqrt[3]{\ }$, $\sqrt[4]{\ }$, etc. In the case of square root, however, the 2 can be omitted (i.e., the signs $\sqrt{\ }$ and $\sqrt[2]{\ }$ both mean square root).

> foot soldiers and 7000 horsemen. The ratio of the stipend was such that 100 *aurei* would hire 18 more foot soldiers than it would mounted men. A certain tribune of soldiers came to the proconsul with 1700 foot men and 200 horsemen and asked for his share of the pay . . .[22]

If the martial setting of this algebra word problem is not to your liking, you can turn your talents to trying to solve an earlier one, posed in the twelfth-century text *Vija-Gan'ita* of Bháskara, Problem 132 (see Colebrooke's *Algebra*):

> The square-root of half the number of a swarm of bees is gone to a shrub of jasmin; and so are eight-ninths of the whole swarm; a female is buzzing to one remaining male; that is humming within a lotus, in which he is confined, having been allured to it by its fragrance at night. Say, lovely woman, the number of bees.[23]

Ineluctably, however, as the sixteenth-century Italian mathematicians allowed particular tactics of solution to particular problems to give way to more general methods applied to more general problems, in their calculations they found themselves nudged more and more urgently, by the momentum of their ideas,[24] to make use of quantities like $\sqrt{-1}$. Especially puzzling is that some of these calculations succeed in giving perfectly comprehensible answers to perfectly comprehensible questions, but only by dealing along the way with somewhat incomprehensible quantities like $\sqrt{-1}$. This can be unsettling; rather like discovering that there is an efficacious way of getting from Brooklyn to Boston, but that somewhere in mid-journey one has to descend to the Underworld.

Here is a concrete example of the type of ordinary-sounding problem that might move a sixteenth-century mathematician to use quantities like $\sqrt{-1}$ to effect its (theoretical, but not practical) solution.

> Suppose that someone has given you the following information about an aquarium tank. The tank holds a volume of 25 cubic feet, and is 1 foot taller than it is wide, and 1 foot longer than it is tall. Find the (precise) dimensions (length, width, height) of the tank.

I said, parenthetically, that quantities like $\sqrt{-1}$ are used to establish a theoretical, not a practical, solution to the problem. To figure out an approximate answer, good enough for any practical considerations about the care and feeding of the fish in the aquarium, there are easier, rougher methods, and even trial and error will do quite well (the aquarium is about $24\frac{1}{2}$ inches wide). The aim here would be to find an exact solution to the problem and, in the course of this, to understand the solution's conceptual structure. You might respond, "What can you possibly mean by the *conceptual structure* of an answer to this problem, which is, after all, a mere number?" Wait.

It was not that such puzzling answers to problems had never been explicitly encountered before then. Nicolas Chuquet, in his 1484 manuscript *Le Triparty*, attempting to find that number whose triple is 4 plus its square, discovers that his method comes up with the "answers" (which I give in modern notation)

$$3/2 + \sqrt{-1.75} \text{ and } 3/2 - \sqrt{-1.75}.$$

And Chuquet concludes that there is no number whose triple is 4 plus its square, because the above answers are, as he puts it, "impossible."[25] This is a perfectly valid conclusion, given that Chuquet was seeking "ordinary number" solutions to his problem. To get a sense, though, of why Chuquet might have been led to think of such expressions—deemed by him impossible—as candidate solutions to the problem, you might try to square $3/2 + \sqrt{-1.75}$ (i.e., multiply this expression by itself using "laws of ordinary arithmetic" plus the fact that the square of $\sqrt{-1.75}$ is $-1.75 = -7/4$), adding 4 to the result, and seeing whether you get 3 times $3/2 + \sqrt{-1.75}$ as the answer.[26]

In contrast to the way in which $\sqrt{-1.75}$ entered as a possible but discarded solution to Chuquet's problem, the novel element in the early Italian involvement with things like $\sqrt{-1.75}$ is that the Italian mathematicians were working on problems having perfectly possible ("ordinary numerical," i.e., real-number) answers, but their methods, at times, involved dealing with numbers like $\sqrt{-1.75}$ along the way.

5. What is a mathematical problem?

Problems are different from questions. We sometimes ask questions in full expectation that the answer will be easily given. "Do you want some more pie?" But we pose (throw out) problems for solution only if we expect that something of a mental stretch is required to come up with the answer.

One can classify categories of straightforward question-asking, as Aristotle does in the *Metaphysics*: "What?" "By what means?" "How?" "Why?" But *problems* are a different story. They seem not to submit easily to any simple categorization. Their posing may take ingenuity:

> *How hadde this cherl ymaginacioun*
> *To shewe swich a probleme to the frere?*

asks the lord in "The Summoner's Tale" in Chaucer's *Canterbury Tales.*[27] *Problems* are the mainstay of the schoolroom, and the melancholy plight of students is that they are bent over their desks working out problems set by others, not by themselves.

All the best mathematical problems are *come-ons*: there is a gentle irony behind them. The problem-setter usually presents to you a very precise task. *Solve this!* An equation, perhaps: just solve it. But if the problem is really good, a solution of it is nothing more than a letter of introduction to a level of interaction with the material that you hadn't achieved before. Solving the problem gets you to a deeper level of question-asking. The problem itself is an invitation, a goad, to extend your imagination. This is true of good school problems but is also true of some—perhaps all—of the famous and venerable mathematical problems. For example, there is the *Poincaré conjecture*, one of the great yet unachieved goals of three-dimensional geometry.[28] The Poincaré conjecture is a precise claim about the characterization of three-dimensional space, and mathematicians would keenly like to know: "Is it true?" "Is it not true?" But the impetus behind the problem is far greater than determining whether it is true or false. Work on the problem presents a possible way of extending our three-dimensional geometric intuition. Now, you might say that we all know three-dimensional space: we get into and out of our sweaters, we tie things together with knots, we dance, we explore caves and mountains. The Poincaré conjecture tells us—plus ultra*—that there is

*Before the discovery of America, *Ne plus ultra* was the motto of the royal arms of Spain, the western limit of the known world. *Beyond us*, proclaimed the motto, *there is no more.* After the discovery, however,

more to be imagined, there are yet ways in which our three-dimensional intuition might be refined, and it challenges us to do so.

---

when Charles V inherited the throne of Aragon and Castille, he simply deleted the *Ne* from the motto: *There is [even] more.*

2

SQUARE ROOTS AND THE IMAGINATION

6. What is a square root?

Thus far we have discussed, for example, the square root of 2 ($\sqrt{2}$), the number whose square is equal to 2, and have seen that $\sqrt{2}$ is also the length of the diagonal of a square whose sides are of length equal to 1. We can give the square root of 2 to any degree of accuracy we wish. Do you want it to ninety-nine decimal places? Here it is:

It was known to the Pythagoreans that $\sqrt{2}$ *cannot* be expressed as a fraction, that is, as a ratio of whole numbers. This, some say, represented a great dilemma to the Pythagoreans, who held that the very building block of the cosmos is Number. For here, one of the most basic proportions in geometry—the ratio of the diagonal of a square to its side—is not expressible as the ratio of one whole number to another. There is a legend that the Pythagorean who revealed this fact to people not of Pythagoras's sect was drowned in punishment.

But since the elegant proof is now fairly widely known, there is no danger, and possibly some delight, in quickly sketching why $\sqrt{2}$ cannot be expressed as a ratio of whole numbers. As arranged below, the proof of this begins with something of a prelude and is then established in four steps.[1] If this type of argument is new to you, it may be more instructive and perhaps more persuasive if you first read the prelude, which will give the strategy of the proof, and think about it before embarking on the steps proper. Each numbered step depends upon the ones preceding it, and each has its aim stated immediately following its head. For each step, try your hand at guessing how to achieve the aim before taking a look at the proof.

Prelude Suppose that $\sqrt{2}$ could be expressed as a ratio of whole numbers. That is, suppose we could write an equation of the form

$$\sqrt{2} = \frac{A}{B}, \qquad (2.1)$$

where A and B are whole numbers. We will see that such an equation leads us to a contradiction, and therefore no such equation can hold. (Which is precisely what we want to prove.)

We can even assume that in our equation, A and B have no common factors, for otherwise we could simplify the fraction A/B by dividing both numerator and denominator by the common factor, giving rise to an equation of the same form as (2.1), only with smaller numerator and denominator. In particular, we can, and will, assume that we have an equation (2.1) for which it is *not* the case that both A and B are even numbers. One of them, A or B, might be even, but both are not. From this we want to get a contradiction.

Step 1 *To make use of the definition of* $\sqrt{2}$*:* Squaring both sides of our equation, we get

$$(\sqrt{2})^2 = \frac{A^2}{B^2} \qquad (2.1)$$

But since the square of $\sqrt{2}$ is 2, we can write this equation as

$$2 = \frac{A^2}{B^2} \qquad (2.2)$$

Step 2 *To show A is even:* Multiplying equation (2.2) by the quantity B^2, we get

$$2B^2 = A^2, \qquad (2.3)$$

from which we see that A^2 is an even number. Now, the square of any odd number is again odd (why?), and therefore, since the square of A is even, A itself must be even.

Step 3 *To show B is even:* By step 2 we see that A is even, so let us express this fact by writing A as twice a whole number, C:

$$A = 2C.$$

Now, substituting $2C$ for A in equation (2.3) we get

$$2B^2 = 4C^2,$$

and dividing both sides by 2 we are left with the puzzling equation

$$B^2 = 2C^2,$$

reminiscent of equation (2.3). But this tells us that B^2 is even, and therefore so is B.

Step 4 *To note the contradiction:* Steps 2 and 3 taken together show that the very assumption that an equation of the form $\sqrt{2} = A/B$ holds leads to an absurd conclusion: that both A and B are even numbers, despite the fact that (assuming $\sqrt{2}$ is expressible as a ratio of whole numbers) we have found an equation $\sqrt{2} = A/B$ for which A and B are not both even (see the prelude). So $\sqrt{2}$ is not expressible as a ratio of whole numbers.

~ ~ ~

A clean argument. But you might find it somewhat eerie to be working in such a never-never land (in which you assume—contrary to fact, as it turns out—that there exists an equation of the form $\sqrt{2} = A/B$); to be working in a terrain, an Erewhon, that allows for rational investigation (given its own rules) and reveals its own impossibility. This argument, troubling as it may have been to the early Pythagoreans, has ushered us into the splendid world of numbers *that cannot be expressed as fractions.*

Splendid world of numbers that cannot be expressed as fractions—what kind of world is that, whose denizens are characterized by a property they do not possess? I am usually impatient with mathematical expositions that end with some stubbornly negative result, if that result can be viewed as the precursor of a deeper positive one. For example, if, as we have just shown, the square root of 2 cannot be expressed as a fraction, the natural next question is, then, how else can you pin it down? Even better would be to give a concise characterization of $\sqrt{2}$ that on the one hand makes it easy to approximate $\sqrt{2}$ as accurately as one might wish, by fractions, and on the other makes it relatively easy to see that $\sqrt{2}$ itself is not a fraction. It would take us too far afield to develop this theme, but I can't help just displaying, if not explaining, a formula that is the seed of at least one answer to this question:[2]

$$\sqrt{2} = 1 + \cfrac{1}{2 + \cfrac{1}{2 + \cfrac{1}{2 + \ldots}}}$$

We have not gotten far in answering the title question of this section, so we need another section with the same title.

7. What *is* a square root?

Any positive number has a square root. A few decades ago, students in at least some primary schools were taught, along with "long division," methods of computing "by hand" the square root of numbers (e.g., the square root of 4.938). Nowadays calculators can be relied upon to take up that kind of chore. You punch any number, say 4.938, into your calculator and press the *square root* key, or, in one computer program, you type in *SQRT* before your number and the answer flashes on the screen:

$$\text{SQRT}\ (4.938) = 2.222\ldots$$

So, whatever it is, "square root" (alias SQRT) is an old friend.

But what is it? More helpful than you might first think is the *tautological* answer to this question:

> The square root of a positive number R is the positive number that is the solution to the equation $X^2 = R$.

Let us denote this positive number $\sqrt{R}$, so we are guaranteed, by its very definition, that

$$(\sqrt{R})^2 = R.$$

But there is also another number whose square is R: the negative of $\sqrt{R}$. For, $-\sqrt{R}$ times $-\sqrt{R}$ is equal to R. Why this is so, or, to put it colloquially, why it is true that *minus times minus is plus*, is a question we will be working on in the next few chapters.

Granting its truth provisionally, then, we may wish to refer to the second solution of the equation $X^2 = R$ as a "square root" of R as well. So for completeness we may say that there are two square roots of R, the *positive* one, which we call $\sqrt{R}$, and the *negative* one, which is just $-\sqrt{R}$:

$$(-\sqrt{R})^2 = R.$$

Tautological though this answer is, it might remind you of the *quadratic formula*, famous in high school algebra classes, and already known to the ancient Babylonians.

8. The quadratic formula.

We have talked about Chuquet's problem (see sect. 4), which we can rephrase as: *find that number whose square is equal to its triple minus four.* If we let X denote the number we want to find, we are trying to solve the equation

$$X^2 = 3X - 4$$

or, equivalently,

$$X^2 - 3X + 4 = 0.$$

We are looking for a numerical quantity to substitute for the unknown, X, to make the value of the expression $X^2 - 3X + 4$ equal to zero.[3] The solution can also be called a *root* of the polynomial* $X^2 - 3X + 4$, for such a numerical quantity would have the desired property: its square is equal to its triple minus 4.

The purported aim of the quadratic formula is to do no less than solve all such problems: *find that number whose square is equal to any specific multiple of itself plus (or minus) any specific number.* Alternatively, we can express this general problem as the problem of solving for the unknown X in the equation

$$X^2 + bX + c = 0;$$

or, equivalently, of finding those quantities (values of X) for which the square of X added to b times X plus c is zero. The b and the c we are to think of as numbers given to us by our particular problem.

The letters early in the alphabet (b, c) are generally

*For a brief review of terms like *root* and *polynomial*, see endnote 3.

taken to be *known* quantities, while the end letters (X, Y) generally stand for *unknowns.** The idea of explicitly differentiating between "not yet specified" values b, c, . . . as if they are known, as distinct from quantities X, Y, . . . which are sought (unknown), already occurs in the writings of the sixteenth-century mathematician François Viète, whom we will encounter again in chapter 4. (He, however, made use of a different alphabetic mnemonic to distinguish between unspecified known quantities and unknowns: the unknowns he designated by uppercase vowels, and the unspecified knowns by consonants.)

You may recall from high school that there are, in general, two values of X that solve the equation

$$X^2 + bX + c = 0.$$

Namely, the *quadratic formula*:

$$X = \frac{-b + \sqrt{b^2 - 4c}}{2} \text{ and } X = \frac{-b - \sqrt{b^2 - 4c}}{2}$$

The Quadratic Formula

*So mathematical practice follows the surprisingly widespread sentiment that the later letters in the alphabet are harder (whatever that means) than the earlier ones. For example, in Virginia Woolf's *To the Lighthouse* where Mr. Ramsay's abstract thinking is likened to a heroic expedition getting further and further into the alphabet: "Z is only reached once by one man in a generation. Still if he could reach R it would be something. Here at least was Q." Consider also Dr. Seuss's *On Beyond Zebra.*[4]

To check that these values of X "work," you just need to be courageous enough to substitute each of these expressions,

$$\frac{-b+\sqrt{b^2-4c}}{2} \text{ and } \frac{-b-\sqrt{b^2-4c}}{2}$$

for X into $X^2 + bX + c$ and confirm that the answer is zero.

For the ancient method of "completion of the square," which allows us to derive the quadratic formula, see the appendix (funny verb: *derive*).

It pays to spend a minute considering what the quadratic formula actually gains for us. It tells us, first, that *if* we can find the square root of numbers (specifically, of $b^2 - 4c$), we can find the solutions of *all* the quadratic equations, $X^2 + bX + c = 0$; and second, that there are, in general, two solutions—two values of the unknown, X, that solve this equation—just as there are two square roots of any positive number!

In fewer words, to solve complicated equations such as $X^2 + bX + c = 0$, we need only solve simpler equations, such as

$$Y^2 = d,$$

for the unknown Y (where d is the number $b^2 - 4c$). We then get X using the recipe in the quadratic formula

$$X = \frac{-b+Y}{2} \quad \text{and} \quad X = \frac{-b-Y}{2}.$$

Of course, this would be of relatively little use if it were an utter mystery how to solve those simpler equations, that is, how to obtain square roots. But happily, we are in good shape here, at least when we are faced with the task of finding the square root of a positive quantity.

To give something of an architectural analogy, think of algebra as a house and the problems that make up algebra as the beams and joists and walls, some load-bearing, some not. The quadratic formula tells us that the problem of finding a square root is indeed a load-bearing one. Solve it, and it will support lots more than its own weight: it will provide the solution to more general quadratic equations. No wonder, then, that finding square roots was viewed as a valuable enterprise and, later, that higher roots were looked to as a possible first step in seeking solutions to more general problems.

9. What kind of thing is the square root of a negative number?

As mentioned in the introduction (sect. 1), if you take any "ordinary" nonzero real number, be it positive or negative, and square it, you get a positive number. So $\sqrt{-1}$ (alias the number, if such a thing could be imagined to exist, whose square is -1) cannot be an "ordinary" quantity. The clear and sober judgment of Chuquet that such objects are "impossible" is echoed everywhere. "We do not perceive any quantity such as that its square is negative!" is a statement made by an Indian commentarist on the twelveth-century text *Vija-Gan'ita.*[5]

Nevertheless, this $\sqrt{-1}$ did not go away. It showed itself as imperatively useful. Those who refused to deal with it did so at the price of limiting their power as algebraists. By the beginning of the eighteenth century, square roots of negative quantities were routinely met with and jauntily handled, and their impossibility noted—as in this statement from a little manual on algebra written by Isaac Newton: "[I]t follows that the Equation has one true Affirmative Root, two negative ones, and two impossible ones."[6]

The story of the ripening act of coming to live with $\sqrt{-1}$, and eventually achieving a somewhat surprising but wholly satisfactory "imagining" of $\sqrt{-1}$, spans more than three centuries of mathematical activity.

10. Girolamo Cardano.

Girolamo Cardano invoked and used "imaginary numbers" like $\sqrt{-1}$. He was born in 1501 in Pavia and died in 1576 in Rome. Cardano wrote on numerous subjects besides mathematics (including medicine and astrology). Even though most of his manuscripts are now lost, his extant works fill ten volumes. His book on gambling, *Liber de Ludo Aleae,*[7] is credited by Persi Diaconis as being the original "invention of the basic approach and rules of calculation for dealing with probability as it appears in games of chance."

With all that, the treatise *Ars Magna* [The Great Art], published in 1545, stands as Cardano's most influential achievement. Here is a translation of the title page:

> THE GREAT ART
> or
> The rules of Algebra
> by
> GIROLAMO CARDANO
>
> *Outstanding Mathematician, Philosopher and Physician In One Book, . . . Which is called the Perfect Work*

Cardano writes in the first chapter:

> Since this art surpasses all human subtlety and the perspicuity of mortal talent and is a truly celestial gift and a very clear test of the capacity of men's minds, whoever applies himself to it will believe that there is nothing that he cannot understand.[8]

There is a shimmering quality to this statement. The "art" is celestial and "surpasses all human subtlety"; it is of pure intellect beyond "mortal talent." And yet Cardano concludes by saying in a wonderful, universally egalitarian manner that "since" all this is so, "whoever applies himself to it will believe that there is nothing that he cannot understand." A peculiar, but to me very appealing, use of the word *since*. It also captures an aspect of mathematics that any teacher of math knows well. On the one hand, mathematics very much lives up to its Greek name (meaning, roughly, "that which can be learned"), for the substance of mathematics can indeed be learned in the most demanding of senses, and all humans can do it, independent of their external experiences, of their language, of their worldly knowledge. Mathematics is, in this sense, available to anyone. On the other hand, it does push mortal talent to its limits.

All math teachers have days that radiate with the optimism of, say, the math lesson in Plato's dialogue *Meno*.[9] These are days when you feel that mathematics is as accessible to everyone as the summer breeze and, better than that, is *already* in everyone's repertoire. But I also suspect that even good math teachers have other days when they wonder whether math "can be taught even to one man, except if he be wise and able to understand by himself" (to pervert the meaning of a phrase from Maimonides' *The Guide of the Perplexed*).[10]

11. Mental tortures.

Cardano, in considering the "quantity" $\sqrt{-9}$, writes: "$\sqrt{-9}$ is neither +3 nor −3 but is some recondite third sort of thing [quaedam tertia natura abscondita]."[11] At one point in *Ars Magna*, Cardano finds himself forced to invoke a square root of −15. He says to the reader, giving no further justification, "You will have to imagine $\sqrt{-15}$," and then he goes on to calculate with it, even though he says that he is doing this by "dismissing mental tortures."[12] The colorful Latin phrase Cardano used for this is *dimissis incruciationibus*, and the translator notes that Cardano might very well be playing on a possible double meaning of this phrase in the sentence, which can be read either as "Dismissing mental tortures, multiply $5 + \sqrt{-15}$ by $5 - \sqrt{-15}$ . . ." or as "Cancelling out cross-multiples, multiply $5 + \sqrt{-15}$ by $5 - \sqrt{-15}$ . . ."[13]

Later on, we too will be multiplying such numbers, but to see what is meant by "canceling out cross-multiples" when you perform the requested multiplication, note that the product of $5 + \sqrt{-15}$ and $5 - \sqrt{-15}$ will be the sum of four terms:

$$5 \times 5$$
$$5 \times (\sqrt{-15})$$
$$5 \times (-\sqrt{-15})$$
$$(\sqrt{-15}) \times (-\sqrt{-15}).$$

The second and third terms, called the cross-multiples, indeed cancel each other out. The first and last terms add together to give 40. Thus, dismissing mental tortures (or not!),

$$(5 + \sqrt{-15}) \times (5 - \sqrt{-15}) = 40.$$

Further on in the exposition of this calculation, Cardano exclaims, "This truly is sophisticated [or sophistic; quae vere est sophistica]." Cardano and his contemporaries would have referred to $\sqrt{-1}$ and to any of its companions—$\sqrt{-2}$, $\sqrt{-3}$, and so forth—when they referred to them at all, as "sophistic negatives," and also as *fictae*. But so were negative numbers referred to as *fictae*. For example, about a certain problem, Cardano writes that "there can be no true solution, but the fictitious one will be -3." It is just that some *fictae* now sit well in our imagination, while others require the stretching of our mental cordons before our imagination can provide a suitable habitation for them. For Cardano, these "truly sophisticated," recondite "third sorts of things," these "sophistic negatives," are useful but perhaps not vital. His followers, Rafael Bombelli and Lodovico Ferrari, however, will find the use of such imaginary numbers essential in the general expression of the solutions to third- and fourth-degree polynomial equations.

In 1572, two years after the second edition of Cardano's *Ars Magna* appeared, Bombelli's extensive treatise on algebraic matters was published.[14] Later we will get a closer look at how Bombelli treats imaginary quantities (e.g., $+\sqrt{-1}$ or $+2\sqrt{-1}$). Since these unruly symbols defy even the simplest of the usual classifications of numbers (like being positive or negative!), Bombelli invents nomenclature to deal with such inconveniences. He refers to $+2\sqrt{-1}$, say, as *più di meno*, which, it seems, is a contraction of *più radice di meno.*[15] It might be translated as "the positive of the negative radical" to distinguish numbers like $+2\sqrt{-1}$ from numbers like $-2\sqrt{-1}$, which Bombelli calls *meno di meno.* And he carefully spells out laws of the form:

"Più di meno via più di meno fa meno"

(e.g., $\sqrt{-1} \times \sqrt{-1} = -1$)

and

"Meno di meno via più di meno fa più"

(e.g., $(-\sqrt{-1}) \times (\sqrt{-1}) = +1$).

3

LOOKING AT NUMBERS

12. The problem of describing how we imagine.

The stretching of the imagination to embrace an otherwise unembraceable fictum would, if Elaine Scarry's premise applied here, be unavailable to us as a felt experience. This would suggest, for example, that no matter how assiduously we study the three-century-long encounter with $\sqrt{-1}$, we will not get any closer to an inner experience of this grand act of imagination-stretching—because there is no inner experience to understand. It would allow only the existence of a *before* (the imaginative act) and an *after*.

There may be a difference between feeling the experience and being able to describe it adequately. Even those who try to articulate, to classify, the fruits of the imagination, and who are committed to the existence of an inner experience concomitant with it, admit to dark difficulty in describing it. The frustration induced by trying to describe the imagination is like—if I may borrow an image of Wittgenstein—trying to repair a torn spider's web with your fingers.[1] Eva Brann calls the imagination the missing mystery of philosophy and an "unacknowledged question mark."[2]

• It is not, however, that people haven't hinted at the motions and inner workings of the imagination. The Stoic Chrysippus posited a kind of elementary particle of the imagination, called a *phantastikon*, an interior analogue of a sensory perception, which impresses itself upon the wax tablets of the mind. The early twelfth-century Sufi Ibn al-'Arabī (in an extensive work, the *al-Futūhāt al-Makkīyah* [The Opening], which has not yet been entirely edited, let alone translated),[3] views the imagination as a *horn of light*, intermediary between "being" (its narrow end) and "nonbeing" (its infinitely wide end).

• Nor is it that we lack sufficient record of the agonies of craft in the service of the imagination. Think of these lines from Yeats's poem "Adam's Curse":

> *I said, "A line will take us hours maybe;*
> *Yet if it does not seem a moment's thought,*
> *Our stitching and unstitching has been naught.*
> *Better go down upon your marrow-bones*
> *And scrub a kitchen pavement, or break stones*
> *Like an old pauper, in all kinds of weather;*

For to articulate sweet sounds together
Is to work harder than all these . . ."[4]

• Nor do we lack close analyses of the process of the imagination. For example, the literary critic John Livingston Lowes scrutinized the notebooks of Coleridge to discover from what materials Kubla Khan's Xanadu was built, which twig by which twig.[5] Lowes's account, though, has quite a different flavor than Coleridge's own version of the creation of "Kubla Khan" in the prefatory note to the poem. After having taken an "anodyne" and, in consequence, having fallen asleep in his chair at the moment he was reading "Here Khan Kubla commanded a palace to be built . . . ," Coleridge claims to have composed at least two to three hundred lines in his sleep. About this, Coleridge comments: "[I]f that indeed can be called composition in which all the images rose up before [me] as *things*, with a parallel production of the correspondent expressions without any sensation or consciousness of effort."[6]

• Nor do we lack reports expressing precise "consciousness of effort." One poet, Stephen Dobyns, describes the act of writing a poem as "leaning against a closed door."* Another contemporary poet writes of the unsettling feel of a "poem about to come": a feeling

*The last essay of Stephen Dobyns's *Best Words, Best Order* (St. Martin's, 1997) is a detailed description of the stages of composition of his poem "Cemetery Nights." He writes poems, he says, by starting with

as disturbing as "the aura of a migraine."[7] "To give birth to an idea" is a metaphor so often used that we no longer wince with sympathetic labor pains when we hear it—even when Rilke uses it: "Everything," he wrote to the poet Franz Kappus, "is gestation and then birthing."[8]

All this may give evidence of the conviction

- that things never seen by corporeal eye do come to the imagination by some means;
- that preparation for this event can be assiduously, if unconsciously, made; and
- that one may sense its onset as one senses, say, a low-pressure front.

But all this gives no clue of how this work is really done.

13. Noetic, imaginary, impossible.

In the early seventeenth century, the mathematician Thomas Harriot, trying to protect numbers like

---

a number of aural, emotional and intellectual concerns floating with a series of images like flies circling in the center of a room. I repeat the rhythms and sounds in my head, run through the images as if through a tray of slides, and lean against the concerns as one might lean against a closed door.

The poem comes when I am suddenly able to join these concerns together under the aegis of one idea or feeling.

$5 + 2\sqrt{-1}$ from reproach, would refer to them as *noeticae radices* ("noetic radicals," or "radicals of the intellect"), and others would refer to them as "imaginary roots," as we continue to do. But even half a century after a geometric rationale for imaginary numbers had been discovered, they were still taxing comprehension. In an 1847 article, the mathematician A. L. Cauchy—whose attitude toward $\sqrt{-1}$ is, briefly, that $\sqrt{-1}$ should be considered something about which we know one salient fact, namely, its square can be evaluated (as −1)—emphasizes that his viewpoint will bring *"la théorie des imaginaires"* within the grasp of all intellects.[9] In 1849, Augustus De Morgan writes:

> The use, which ought to have been called experimental, of the symbol $\sqrt{-1}$, under the name of an impossible quantity, shewed that come how it might, the intelligible results (when such things occurred) of the experiment were always true, and otherwise demonstrable. I am now going to try some new experiments . . .

De Morgan, in reviewing the acceptance of imaginary numbers, says that insofar as they have "no existence as . . . quantity," they are "permitted, by definition, to have an existence of another kind, into which no particular inquiry was made."[10]

I have quoted Cauchy and De Morgan specifically to

remind us that the idea whose birth pains we hope to reexperience did not relieve all difficulties of comprehension connected to the imaginary quantities.

14. Seeing and squinting.

I said earlier that Scarry's claim would allow only the awareness of a *before* (the imaginative act) and an *after*. But can we catch some intermediate moment in the act of imagining Ashbery's phrase "the yellow of the tulip"? Does our mind's eye shuttle, for example, between the words *yellow* and *tulip*, each visit to one of these words intensifying the effect of the other? The "yellow" in this phrase dazzles all the more because it is "of the tulip," while the tulip blooms fresher in our minds because of its yellow flame. Perhaps one might chart this mind motion the way psychologists of vision chart our eye movements as we contemplate a painting; for example, Van Gogh's *Sunflowers*.

On the face of it, the tulip, in that phrase of Ashbery's, is a means and not an end. The tulip makes its appearance only so that we might all the more easily conjure up that precise yellow, which happens to be its color; it is only the *color* that we are explicitly invited to imagine. But then, in its turn, the dutiful servant to the phrase, the tulip, having summoned up such a fine yellow, becomes the focus of our fickle attention.

But there is also an inherent motion in the mere act

of thinking about the color yellow itself, a color that both attracts our heliotropic eye and, at the same instant, dazzles it.

As I flip through the fall 1997 White Flower Farm catalog, I notice that, despite the title, it is really the yellow flowers in the book that capture me. And in its description of the tulip called "Sweetheart," the catalog assures me that "the eye reads it as pure sunshine." This is true, and as with looking at the sun, you can't do it for too long: you instinctively soon shift your gaze from the picture, only to find that the yellow patch turns to its opposite, its negative: a purple tulip. This phenomenon is quite unmistakable when it happens. I'm also told that yellow flowers have somewhat purple shadows, but I have never been able to discern this myself. Does double negation on the color wheel get you back to where you have started? That is, can you also detect some yellow in the shadow of a violet?

15. Double negatives.

If you have not done arithmetic in a while, you might have been puzzled when I said (in sect. 7) that the product of a negative number and a negative number is positive. But arithmeticians agree with grammarians in claiming that a double negative is a positive. A suddenly forgiven debt is, in effect (when canceled in our account books), a newly acquired asset. Perhaps we should come to terms with this.

Negative numbers, those peculiar quantities invented by Indian mathematicians over a millennium ago, are, of course, "symbolized debits." If I owe 5 dollars to someone, I set down in my ledger book: -5. If I owe 5 dollars to each of three different creditors, I set down

$$3 \times (-5) = -15$$

If I owe 5 dollar bills to no one, I am entitled to write

$$0 \times (-5) = 0.$$

But allow me to be a mathematician as well as a somewhat more profligate debtor, and suppose that I owe 5 dollar bills to each of N creditors. I must sadly set down my debt as

$$N \times (-5)$$

and hope for some improvement of my assets. If I gain 5 dollars, I can now pay off one person, and when I do, I will owe 5 dollars to one less person. So, I now have only $N - 1$ creditors, and I will record this transaction by writing

$$N \times (-5) + 5 = (N - 1) \times (-5). \qquad (3.1)$$

Or, if you allow me to move the lone 5 on the left-hand side of this curious equation to the right-hand side, changing (of course) its sign, I could write

$$N \times (-5) = (N - 1) \times (-5) + 1 \times (-5), \quad (3.2)$$

which seems, after all, fair enough: $N - 1$ "negative fives" plus one more "negative five" gives N "negative fives." As written above, this line in my ledger is only a template record of any one of a number of possible transactions, depending, of course, on N (the number of creditors I initially had). It is clear what equation (3.1) would mean if, for example, N were 3. It would read, in slightly expanded form,

$$3 \times (-5) + 5 = (3 - 1) \times (-5) = 2 \times (-5) = -10.$$

Or, if N were 4, it would read

$$4 \times (-5) + 5 = (4 - 1) \times (-5) = 3 \times (-5) = -15.$$

But all this accounting of paid-off debts is premised on the assumption that I actually have creditors and therefore have debts to pay off. Or does equation (3.1) also make sense even if I have no creditors? In that case, initially I would take N to be zero and my template record—which, in its naive interpretation, would seem to be involving (on the right-hand side of the equation) a negative number of creditors, whatever that means—would read

$$0 \times (-5) + 5 = (0 - 1) \times (-5).$$

Working out the left-hand side of this equation, since we have already understood $0 \times (-5)$ to be 0, we get the tally $0 \times (-5) + 5 = +5$, while the right-hand side is visibly $(-1) \times (-5)$. So,

$$+5 = (-1) \times (-5),$$

which would seem to be telling us that, indeed, negative times negative is positive.

But how can we justify the arithmetic acts we have gone through? We have depended, it would seem, on the truth of equation (3.1) or, equivalently, equation (3.2) when we substitute zero for N. On what grounds have we the "permission" to perform this calculation? Can we construct a mental image of what we have done? Is our conclusion, $+5 = (-1) \times (-5)$, "correct"? And equally important, if it is correct, what does it mean to say that it is correct?

If you are bewildered by this series of questions, you can understand the distress of fourteen-year-old Marie-Henri Beyle (later known through his writings as Stendhal), who, on quizzing teachers and friends about "minus times minus is plus," found no one who could justify it to him. In his fragmentary and vinegary "autobiography," *The Life of Henry Brulard*, Stendhal muses that his early enthusiasm for mathematics may have been based on his loathing for hypocrisy, by which he means a loathing of some of his relatives and some of their priests.[11] "Hypocrisy was impossible in mathematics," he writes. But then,

> What a shock for me to discover that nobody could explain to me how it happened that: minus multiplied by minus equals plus (– × – = +)! . . . Not only

> did people not explain this difficulty to me (and it is surely explainable, since it leads to truth) but what is worse, they explained it on grounds which were evidently far from clear to themselves.

His teacher, rather than responding to his objections,

> grew confused . . . and eventually seemed to tell me: "But it's the custom: everybody accepts this explanation. Why Euler and Lagrange, who presumably were as good as you are, accepted it! . . . It seems you want to draw attention to yourself."

Stendhal then turned to the brilliant students in his class with his question; they laughed. He wrote: "Can my beloved mathematics be a fraud? I didn't know how to reach the truth. Oh how eagerly I would have . . . I might have become a different man; I should have had a far better mind."

Let us take a short break from "minus times minus" and from equations (3.1) and (3.2), but let us not forget Stendhal's lament. We shall return.

16. Are tulips yellow?

Some are. Their conjectured progenitors among the wildflowers originating in the Asian steppes were claret red, yellow, pink, or white; "Sometimes different colors merge imperceptibly in the same flower."[12] But when they were brought to the Middle East in the tenth and eleventh centuries, their cultivated colors included blood red. There is the Persian tale of a prince named Farhad, who in Romeo-and-Juliet fashion erroneously thought his lover, Shirin, had died. In grief, he axed himself to death, and where each drop of his blood fell to the ground a scarlet tulip bloomed.[13] In the sixteenth century, the tulip had become, under Süleyman the Magnificent, a major motif of Ottoman artists and artisans—it was being woven into the velvet coverings of saddles and into the prayer rugs sewn by brides, and rows of tulips were embroidered in Süleyman's "gowns of cream satin brocade."[14] The tulip was then being bred in new varieties, with the flares of color that would eventually capture the imagination and the pocketbook of the Dutch.

17. Words, things, pictures.

Maybe the shock of the image evoked by "the yellow of the tulip" comes from the simplicity of the *shape* of a tulip. Its stalk is so straight, so smoothly purposeful. The calyx and the chalice of its petals can be turban-tight. The most prized Istanbul tulips had petals that were almondlike: "long, slender, and needle-pointed at the tip."[15] "In a perfect flower the petal would conceal the stamens . . . The flower had to stand erect on its stem, thin and well-balanced."[16] The tulip has the innocent poise of a closed lotus.

For the Ottomans, the tendency for the flower,

when in full bloom, to bow its head made it a representative of the virtue of "modesty before God."[17] It was called the "flower of God," because the Arabic letters making up its name (*lale*) are the same as those that spell *Allah*.

In contrast, "Here tulips bloom as they are told," wrote Rupert Brooke.[18] This line conjures up for me the picture of a flower as a child would draw it, with the connection between the word *tulip* and the thing *tulip* as tight as the connection between *words* and *pictures* in the books (or boards) that teach children to read. I have always been fascinated by the books and other devices that teach children to read,* and I particularly love one such device, a wooden board of the sort traditionally used in the Netherlands to teach reading, one of which a friend once gave to me. These "reading boards" are called *leesplankjes*.

On the *leesplankje* appear seventeen simple colored pictures of *things*, and under each picture the corresponding *word* is printed in trim lowercase letters. There is something strikingly unadorned and archetypal about each of the seventeen pictures. A dove, for example, perched on the red tile of a roof, presents it-

*Keats wrote: "I will call the *world* a School instituted for the purpose of teaching little children to read—I will call the *human heart* the *horn-book* used in that School." Letter (1819) to his brother George and sister-in-law Georgiana in Kentucky, in *Letters of John Keats*, ed. Robert Gittings (Oxford Univ. Press, 1970).

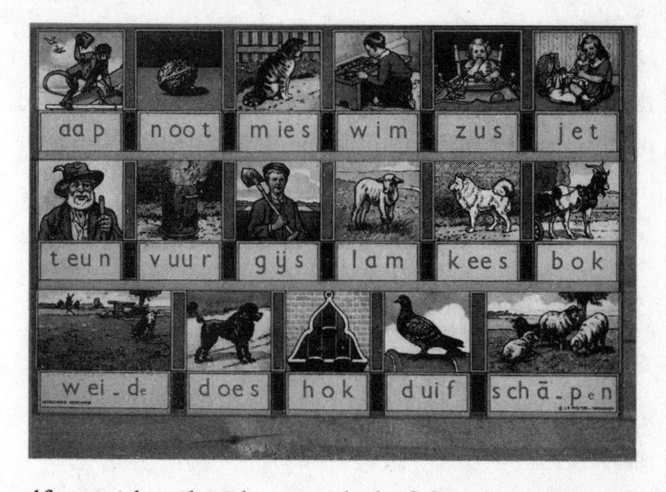

self as neither the Platonic ideal of doveness nor as *this* dove or *that* dove, but rather as Dove, guileless dove, the Adam of dovedom before the Fall. And under this picture, written neatly, is the single word *duif.* The traditional board has pictures of a goat (*bok*) hitched to a cart, a shorn lamb (*lam*), and a young boy (*wim*), an "everychild" in shorts, kneeling, playing with his *leesplankje*. No picture of a tulip appears on my Dutch *leesplankje*.

But the tulip is, at least for me, the primordial "*leesplankje* flower." Its form, axially symmetric, presents itself to us with utter clarity. Its color commands the kind of vividness evoked by Rimbaud's strident assignment of colors to vowels, "A *noir*, E *blanc*, I *rouge*, U *vert*, O *bleu: voyelles*" (which, apparently, repeats the coloring scheme of letters in the standard French primer of his time).[19]

18. Picturing numbers on lines.

Perhaps we still don't know why the product of two negative numbers is positive, but let us draw that ledger book of ledger books, the *number line*:

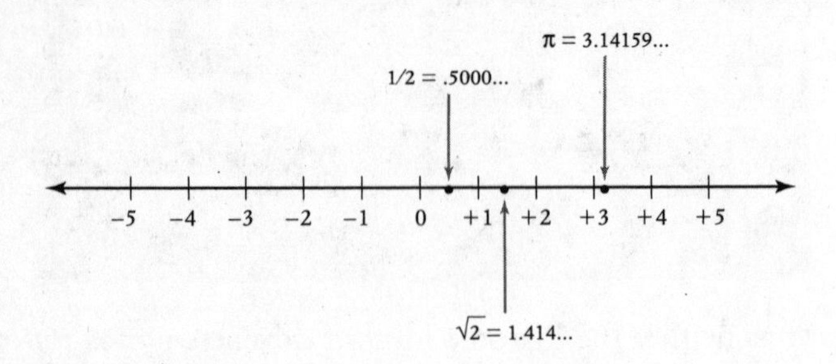

The number line is usually drawn horizontally, going to infinity both to the left and to the right. The points of the line are identified with real numbers, increasing from left to right, with zero at the center ("the two-faced mirror of zero that can endow / the integers with negative looks at themselves").* As I briefly mentioned in the preface, *real numbers* can be given by their decimal expansions. If a real number is negative, it comes along with a minus sign. For example,

*This is from John Hollander's poem "There or Then" (in his *Figurehead* [Knopf, 1999]), which begins:

> At home, at noon, I am located by three *where*
> Coordinates and one for *when* but none
> For *late* or *soon* which seems
> Unfair: the realm of here and there
> Scorns the immense expanse of now and then
> With its symmetrical maps: . . .

$$-\pi = -3.14159\ldots$$

Positive real numbers that correspond to *finite* decimals (e.g., 2.5) have some alternative decimal expressions (e.g., 2.5 = 2.50 = 2.500 = 2.5000 . . . = 2.4999 . . .). Real numbers that do not correspond to finite decimals have a unique (infinite) decimal expansion. If you know the decimal expansion of two numbers, you can say which of the two is larger, which smaller, and they fit in a single continuous line such as the one schematized in the diagram above, each point of which corresponds to a unique *real number.* I wonder who was the "first" to have drawn such a number line, replete with positives and negatives. What a useful diagram to hang numbers on, and what a familiar one to us. No Greek drew such a marked line in classical times, apparently; no Roman either. But, implicitly, every lever balanced on a fulcrum, subject to Archimedes' law of the lever, is, functionally, like an unmarked number line.

Archimedes' *law of the lever* tells us, for example, the intuitively evident truth that if an adult and a child are seated on opposite sides of a seesaw, and the adult is twice the weight of the child, then, to keep the seesaw perfectly balanced, the child should sit twice as far from the fulcrum as the adult.

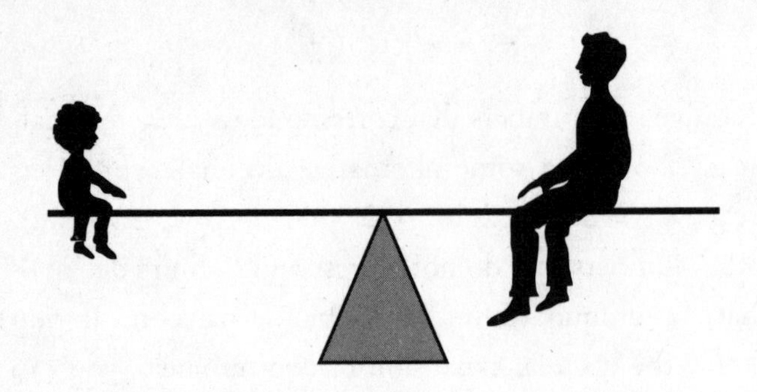

These days, numbers score every thermometer in every nursery. Number lines, marked in gauges, are ubiquitous, decorating dials, meters, quadrants, sextants, compasses.[20] How early in the surviving literature can we find the directional implication: "left" for negative, "right" for positive? In the writings of Brahmagupta in the seventh century, as in the *Vija-Gan'ita* of Bháskara, we find the rules for addition and subtraction of positive and negative numbers (e.g., "a negative minus a positive is a negative," as in $(-3) - 2 = -5$). A sixteenth-century commentator explains Bháskara's calculational rules as follows:

> [H]ere negation is . . . contrariety . . . that is to say, in the contrary direction. As the west is contrary of east; and the south the converse of north. Thus, of two countries, east and west, if one be taken as positive, the other is relatively negative. So when motion to the east is assumed to be positive, if a planet's motion be westward, then the number of degrees equivalent to the planet's motion is negative.[21]

Regarding such directionality, see Barbara Tversky's "Cognitive Origins of Graphic Productions," which studies the question of how children of different cultures orient their maps and drawings that depict increasing quantities.[22] In tabulating children's efforts at organizing information spatially, Tversky's data corroborate our expectations: if encouraged to produce a pictorial chart of increasing numerical values or of flow of time, English-speaking children make their number values, or time, increase as one reads from left to right, or possibly from bottom to top. Arabic-speaking children draw things going the other way, from right to left, and also, possibly, from bottom to top. Neither English speakers nor Arabic speakers will draw things to show an increase as one proceeds from top to bottom. In contrast to what one may consider the approximate neutrality of right-left direction in the horizontal dimension, the psychologists Mark Johnson and George Lakoff emphasize the non-neutrality of the vertical with regard to issues of quantity (and Tversky does the same for issues of emotion).[23] Johnson and Lakoff point out the universality of *more is up* and *less is down.* They offer the examples in English such as: *prices rose, the Dow hit bottom, turn up the thermostat*; and they note that this bias occurs in many languages. Even more telling is that there is no language for which the opposite is true: none in which more is down and less is up. To be sure, we have no end of experience, from childhood on, that helps to entrench the metaphor *more is up*; for example, every time we pour water into a glass the level goes up.

Regarding emotional non-neutrality, Tversky points out that spatial expressions like "I'm sitting on top of the world" and "he's sunk into a deep depression" seem to occur in languages all over the world. The word *deep*, however, may be difficult to pin down in terms of when it is used approvingly and when disapprovingly. But Tversky points out that association of *up* with positivity occurs not just in language but in gesture as well (thumbs up versus thumbs down).

In Indian texts, one seems to find hints of a (pre-Napierian,[24] as it were) "slide-rule" type of motion picture for addition and subtraction: if you want to subtract 2 from a number X, just put your finger at the point X on the number line and slide it along two units to the left; your finger will then be at the point $X - 2$.

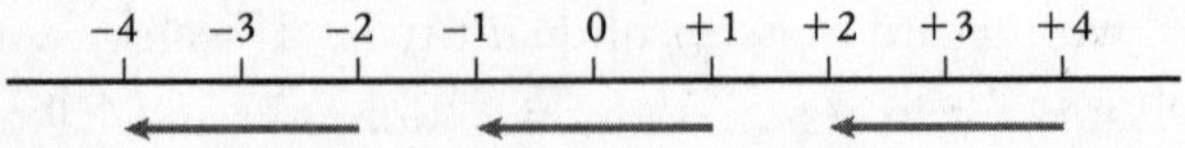

−4 −3 −2 −1 0 +1 +2 +3 +4

It may be useful simply to repeat what we have just done, paying attention to the ever-so-slight, yet ever-so-important change of viewpoint that has been effected. We are explaining subtraction by thinking of the notion of subtraction in *geometric* terms. We envision the operation of subtracting 2 (from various numbers N) in terms of the motion: slide the number line two units to the left.

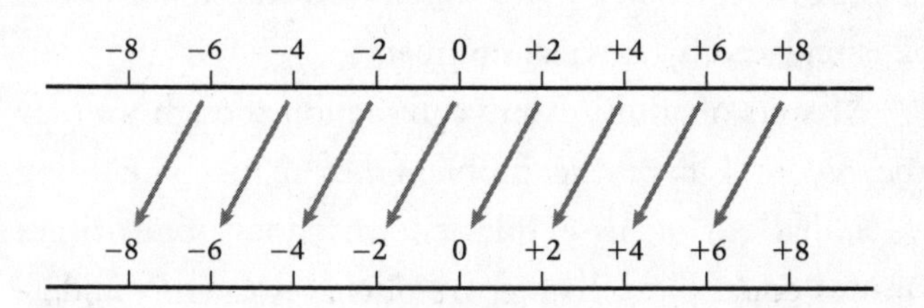

Is it the prevalence of such kinds of number picture that makes understanding computations with negative numbers so easy for modern children, computations that would have been unspeakably difficult for, say, Julius Caesar? A friend of mine was recently teaching subtraction to a five-year-old.[25] "What is 8 take away 6?" he asked. The child answered, "2." "What is 8 take away 8?" The child answered, "Nothing." "What is 6 take away 8? "Nothing, again," was the answer. Which is fair enough. But at that point, my friend simply *drew* the number line for the child, with hardly any further words of explanation, and asked his series of questions again, this time tracing the activity on the line. When he came to the question "What is 6 take away 8?" the answer, unhesitatingly, was "−2." What changed in the imagination of the child?

From here on I will assume that we have all surpassed the fictional Julius Caesar I have alluded to, and that we are utter masters of addition and subtraction, and are happy with the basics for these operations. To add or subtract by 2, you slide the number line to the right or to the left by two units. And that holds also if 2 is replaced by any real number.

Masters of addition and subtraction though we may be, we still have the problem before us of guiding Stendhal out of his perplexity: why does minus times minus equal plus? To prepare ourselves for this undertaking, we will first try to get more familiar with the number line, and then ponder the operation of multiplication.

19. Real numbers and sophists.

In Plato's dialogue *The Sophist*, there is a conversation between someone referred to as "the Stranger" and young Theaetetus.[26] As one knows from the earlier dialogue *Theaetetus*, this is the mathematician who proved that the square root of a whole number cannot be given as a fraction, that is, as a *ratio* of whole numbers, if that number is not a square of a whole number. Thus, $\sqrt{2}$, $\sqrt{3}$, $\sqrt{5}$, and so forth, are not expressible as ratios of whole numbers. (Recall that in section 6 we proved that $\sqrt{2}$ is not expressible as such a ratio.)

In a part of their conversation, the Stranger engages Theaetetus in the task of pinning down the nature of "the sophist." The Stranger proposes to do this by making a series of distinctions that refine and further refine categories, nets of finer and finer mesh, in the hope of capturing the categorical definition of "the sophist." As readers of that dialogue know, the sophist, if he or she exists, eludes them.

The Stranger, I imagine, sensitive to the fact that his conversant is a mathematician, is partly playing on (and with) the (then current) mathematical idea of prescribing geometric proportions by closing in on them from above and below with proportions that are ratios of whole numbers; that is, by capturing these proportions (such as the ratio of the diagonal to the side of a square) in finer and finer meshes describable in the vocabulary of whole numbers.[27]

Think, for example, of capturing a proportion like $\sqrt{2}$:1 by first corralling it between two ratios of whole numbers, one of these ratios being larger than $\sqrt{2}$:1, the other smaller. Then corralling it again, and then again, with corrals of smaller and smaller width.

If you wonder where we are going to get all these corrals, you may be relieved to discover that *decimal expansions* do this work for us very handily. Take the decimal expansion of $\sqrt{2}$ (e.g., as given in the snakelike configuration at the beginning of chapter 2). The first few digits are 1.414 . . . , and this already tells us that $\sqrt{2}$ is wedged between 1 and 2; between 14 tenths and 15 tenths; and more finely, between 141 hundredths and 142 hundredths; and even more finely, between 1414 thousandths and 1415 thousandths, and so on.

Decimal expansions, then, provide "capturing corrals" for real numbers; or, to drop the rodeo imagery, decimal expansions provide systems of *nested intervals* on the real number line—*calipers*, if you wish, delicately poised on the real line, their needles resting on fractions, squeezing more and more tightly about the real number they are designed to capture.

But decimal expansions are not the only way to do this: any such system of calipers that close in on your number would serve to "delineate" it.[28] Here is a real line with some nested intervals converging to the point 0.90625 . . . :

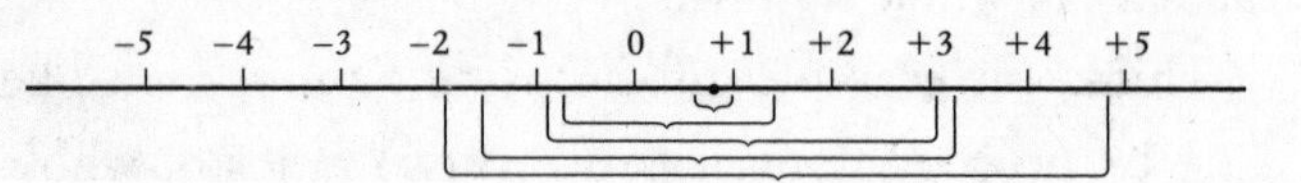

There is a converse to this: any such system of calipers poised on the real line, squeezing in, more and more tightly, will succeed in capturing *some* real number.[29] Real numbers, then, are not as elusive as is "the sophist" in Plato's dialogue.

4

PERMISSION AND LAWS

20. Permission.

Earlier I asked, regarding an arithmetical calculation, whether we had "permission" to perform it (see sect. 15).

As I understand it, the first time Gabriel García Márquez opened Kafka's *The Metamorphosis*, he was a teenager, reclining on a couch. Upon reading

> As Gregor Samsa awoke one morning from uneasy dreams he found himself transformed in his bed into a gigantic insect . . .

García Márquez fell off his couch, astonished by the revelation that you were *allowed* to write like that![1]

Girolamo Cardano, in dealing with powers of quantities, writes:

> For as *positio* [the first power] refers to a line, *quadratum* [the square] to a surface, and *cubum* [the cube] to a solid body, it would be very foolish for us to go beyond this point. Nature does not *permit* it. [Emphasis added][2]

What is this agency, "Nature," and how does it grant or deny permission?

Permitted or not, and foolish or not, Diophantus[3] had already acknowledged fourth powers (viewed as the square squared) and sixth powers (viewed as the cube squared). So did the Hindu mathematicians of the twelfth century, but perhaps what had made such higher powers more imaginable to them was that they were conceiving of their problems exclusively in terms of whole-number solutions. Cardano also has a bit of traffic with higher powers, fifth powers and seventh powers particularly. As he puts it, he deals with these "either by necessity, or out of curiosity."[4]

One problem with *higher powers*, as you can see from the first of Cardano's statements quoted above, is that the powers are directly tied to geometry—the square is conceived as *area*, the cube as *volume*; as a result, the fifth power, for example, seems to hit a geometric impasse. The desire to tie these powers to geometry is understandable, for how else can such powers be visualized? There is also the touchy question of "units": an area might be measured in "square inches" and a volume in "cubic inches." How, then, to measure fourth powers, and, even dicier, how to compare these different "species" in a single formula? With this in mind, it is amusing to turn to the twelfth-century Sanskrit texts on the subject.[5] Here, since it is "whole-number solutions" that are desired, the complex issue of units, and the possible confusion of species of units in the same equation, is simply not regarded as an obstacle. It is treated with somewhat laconic equanimity, as in the word-problem of Bháskara quoted earlier (see sect. 4), which blithely deals with "the square-root of half the number of a swarm of bees."

Here is another of Bháskara's word-problems:

> Of a flock of geese, ten times the square-root of the number departed for the *Mánasa* lake, on the approach of a cloud: an eighth part went to a forest of *St'halapadminís*: three couples were engaged in sport, on the water abounding with the delicate fibers of the lotus. Tell, dear girl, the whole number of the flock.[6]

If you reexpress this as the problem of finding solutions to an equation (as described in note 3), you are looking only for solutions of the equation that can be interpreted as a "number of geese"—that is, you are interested only in positive whole-number solutions. If your equation *also* allows some other solution, which is, say, a fraction but not a whole number, or is a negative number, you may discard it as an artifact of your method of solution—for Bháskara's problems ask for the number of bumblebees in a certain swarm, or swans in a lake, or monkeys, or lotus flowers. In contrast, Cardano, Ferrari, and Bombelli are specifically not restricting themselves to whole-number solutions to the problems they work on: they want to find all real-number solutions.

Four decades after the first publication of *Ars Magna*, François Viète's *Introduction to the Analytic Art* appeared.[7] One of the great advances in Viète's treatise is his proclamation that we can deal with laws of operation of equations which

> no longer limits its reasoning to numbers, a shortcoming of the old analysts, but works with a newly discovered symbolic logistic [per logisticem sub specie] which is far more fruitful and powerful than the numerical logistic for comparing magnitudes with one another.[8]

He formulates principles of operation with unknowns —with *unspecified numbers* ("species,"[9] in his terminology)—the same principles that govern operations with specific numbers. No demonstration of any of the governing rules of his "art" is given. And there are many such rules. For example, he allows us to strike out a common term in an equation to get a simpler equation, passing, say, from $A^3 + BA^2 = C^2A$ to $A^2 + BA = C^2$, asserting the law "An equation is not changed by depression."[10] He proclaims these principles by fiat and ends the introduction to his treatise by saying that his analytic art

> appropriates to itself by right the proud problem of problems, which is:
>
> TO LEAVE NO PROBLEM UNSOLVED.[11]

It has happened to me often, and surely a similar thing happens to all mathematicians, that upon hearing of someone's wonderful new idea, or new construction, I have, like García Márquez, fallen off my (figurative) couch, thinking in amazement, "I didn't realize we were *allowed* to do that!"

I think the issue of "permission" lurks behind many conversations I have heard about the question of "invention versus discovery" in mathematics. The word *permission* is awkward, carries irrelevant associations, and is philosophically unsound (e.g., *who* or *what* is doing the permitting, and *when* do you know that you are granted this permission?). Yet some such term is needed, and I'm still groping for an adequate substitute.

This issue may have significance beyond invention in mathematics. The way that "permission" and potent restriction operate in mathematics has served some as a paradigm for thinking about intertwining freedoms and restrictions in other areas of imaginative work. For example, the philosopher Johann Fichte, with mathematics in mind, generalizes as follows:

> We must freely construct, produce in the imagination, as we did above in the case of the triangle. In this case an evidence will take hold of us, namely that it is only possible *in this way*: a power and thus a law will give shape to our free construction.[12]

In the imaginative world, there is an ever unsettling confusion about whether we are bound in nutshells or can count ourselves kings of infinite spaces. On the days when the world of mathematics seems unpermissive, with its gem-hard exigencies, we all become fervid Platonists (mathematical objects are "out there," waiting to be discovered—or not) and mathematics is all *discovery*. And on days when we see someone who, Viète-like, seemingly by willpower alone, extends the range of our mathematical intuition, the freeness and open permissiveness of mathematical invention dazzle us, and mathematics is all *invention*.

21. Forced conventions, or definitions?

We have left a basic question dangling: why does minus times minus equal plus? For example, is our equation in section 15,

$$(-1) \times (-5) = +5,$$

"correct"? And equally important, if it is correct, what does it mean to say that it is correct?

Let us recall how we came to this equation. We started with the "template ledger book entry," designated equation (3.2),

$$N \times (-5) = (N-1) \times (-5) + 1 \times (-5),$$

and dared to make use of this equation when N is zero. On the face of it, this equation seems harmless enough: $(N-1)$ "negative fives" plus one "negative five" makes N "negative fives." This would seem to hold not only for "negative five" but also for any quantity, C:

$(N-1)$ C's plus one more C makes N C's.

More generally, if A and B are any two counting numbers,

A C's plus B C's makes $(A + B)$ C's.

In terms of an equation,

$$A \times C + B \times C = (A + B) \times C;$$

and what we did in section 15 was to depend upon this equation being true, not simply when A and B are counting numbers, but also when A is negative (in our application, A was -1 and B was $+1$).

This general rule,

$$A \times C + B \times C = (A + B) \times C, \qquad (4.1)$$

is called the *distributive law.*[13] Our question, then, is simply this: Is the distributive law valid if A is -1 and B is $+1$? For this is precisely what we used in section 15 to get "minus times minus is plus."

Let us accept the fact, at least for the moment, that the distributive law for *positive* real numbers is a correct rule and that it expresses an essential relationship between addition and multiplication. Does our acceptance of the distributive law for positive real numbers A, B, and C help us legitimize the validity of the distributive law in the more extended range of all real numbers A, B, and C (positive or negative) as well? Does it oblige us to accept the truth of the distributive law in this extended range? And if we are not *obliged* to do so, are we permitted to do so anyway? And if we are *permitted*, should we? And if we *do*, what are we doing? Are we accepting a *convention* that artfully helps to organize our thought but is nevertheless only one among many competing, and arbitrary, conventions? Or are we accepting a convention that is forced on us by the nature of things? Are we, in accepting this law, conjointly engaged in the act of extending the *very definition* of the operation of multiplication—from the range of positive numbers to that of all numbers, positive and negative?

To help us think about this, perhaps we should backtrack and ask another question.

22. What kind of "law" is the distributive law?

Clearly, the *distributive law* (which for the moment seems to be hung up in committee in this book) is quite a different kind of thing from Newton's laws, or the Corn Laws. In the natural sciences, a *law* is, according to the *Oxford English Dictionary*,

> A theoretical principle deduced from particular facts, applicable to a defined group or class of phenomena, and expressible by the statement that a particular phenomenon always occurs if certain conditions be present.

In mathematics, however, one can distinguish at least two distinct flavors of law. The first kind of law has the comfortable feature that all of its vocabulary has previously been defined; the law makes a particular claim, applicable to a specific class of instances. To be sure, we would hesitate to call it a law unless the claim was proved for that specific class of instances. For example, the statement that "the square of any whole number is not smaller than the number" is perhaps too modest a statement to be referred to as a law. But if we called it a law, we would most likely be thinking of it as a law of this first kind, for we would (most likely) have previously understood and formulated useful definitions of the ingredient concepts: *whole number*, *square*, *smaller*.

The second kind of law has the opposite feature: it is

not the case that all of its ingredient concepts have been previously pinned down; rather, the law itself is what defines (or constrains in some essential way) the concepts contained in its wording. These laws are often referred to as *postulates* or *axioms*. If *Euclid's First Postulate* (the dictum in Euclid's *Elements* that allows us to conclude that "one and only one line passes through any two distinct points in the plane")[14] were considered a "law," it would be a law of this second kind. Although Euclid had provided us with prior definitions of *point* and *line*,* it is the First Postulate that begins to tell us what sorts of things points and lines are.

Given this dichotomy, which kind of law is the distributive law? We will return to this.

The dictionary writers began their definition of *law* in the natural sciences with the phrase "theoretical principle." Returning to the *Oxford English Dictionary*, we learn that a *principle* is

> A fundamental truth or proposition, on which many others depend; a primary truth comprehending, or forming the basis of, various subordinate truths . . .

This underscores two points, both rather evident but both important:

*"Definition 1. A *point* is that which has no part. Definition 2. A *line* is breadthless length." Book I of Euclid's *Elements*; cf. Heath, *Euclid's Elements*, vol. 1, p. 153.

1. A *law* or *principle* is meant to be, in some way, "fundamental," "primary," and "forming the basis of" its subordinate truths.

2. A *law* or *principle* is a template to be applied to "many" other truths, "various" subordinate truths.

I take the first point as reminding us that to call something a law is not a neutral act: you usually use the label *law* only for assertions that you feel have some intrinsic structural priority over their consequences. I take the second point as emphasizing (requiring, even) the *economy* of thought inherent in something we call *law*: a single principle can cover much territory.

The credo of *rationality* is that from a small number of evident principles, many things may be seen to follow. Therefore it is no surprise that economy of expression is highly valued in mathematics, as in the natural sciences. The glory of the calculus lies in the succinctness of its language and in the range and power of its applications as well as in its accuracy. Readers of Ernst Mach's *The Science of Mechanics*[15] discover that the economy of expression of the laws of physics, which condense the outcomes of that science's long experimental tradition, is the very thing that constitutes, at least for Mach, the force of those laws. (This, of course, is an extreme view.)

In poetry, also, conciseness of expression is vital—but, I believe, for quite different reasons. Poetry does its work utterly differently. Think, for a moment, of W. S. Merwin's poem "Elegy,"[16] which is, in its entirety, a string of six one-syllable words:

Who would I show it to

Now, its title proclaims its form, the most common elegiac meters being dactylic hexameter or dactylic pentameter, with each line composed of a string of six, or five, dactyls. (A dactyl is a three-syllable foot of the shape ¯ ˘ ˘; that is, the first syllable is stressed and the remaining two are unstressed, as in the word *funeral.*) Merwin's poem, a single line of *dactylic dimeter* (two dactyls), establishes its pattern and conveys its grief in the briefest manner: its only subject ("I") is an unstressed syllable, as befits a mourner, and its preponderant sound comes from the hollowest of vowels. Surely "conciseness of expression" isn't at all the right phrase for the desperate constriction of this poem, but the poem's concentration of meaning is one source of its power.

5

ECONOMY OF EXPRESSION

23. Charting the plane.

For New Yorkers, the mention of Eighth Avenue and Forty-second Street calls to mind a single intersection. This concise way of denoting position (by giving two numbers: 8 and 42 in this case) is like our system of latitude and longitude for pinpointing spots on the globe. Nowadays, with the Global Positioning System (GPS) working so well, if we wish to get to any place, say a specific bobbing buoy in the Chesapeake Bay during a deep fog, we need only know two numbers: the latitude and longitude of the buoy. We sail our craft so that the GPS readout tells us that we have reached that latitude and longitude, then we peer over the bow. A pair of numbers will get us there.

There is an alternative way of fixing a position. If I say, "Five yards north-northwest of the oak tree lies the buried treasure," and if you know where that oak tree is, and if I am telling the truth, you will find the treasure. Here we have a known reference point (the oak tree), and we have stipulated a distance from it, and a direction. This is a close kin of the radar operator's mode of positioning, where the known reference point is his or her own position (let us call it "home") and the position of the object the operator is interested in is determined graphically in terms of distance and direction from home. As with latitude and longitude, to pin down a position by this method—assuming you know where home is—you must give two pieces of information: distance and direction.

Both ways of determining positions on our planet have their mathematical counterparts in methods of determining—or "naming"—points on the (infinite) Euclidean plane—that is, the plane studied in high school geometry: that plane on which straight lines are extended, angles are bisected, and ruler-and-compass constructions are made.

The "latitude and longitude" way of describing points on the Euclidean plane corresponds (very roughly) to an idea introduced by René Descartes (1596–1650), usually called *Cartesian coordinates*, thanks to which, as Descartes comments, the relationship between algebra and geometry is "expressed in characters in the briefest possible way." He adds: "I would borrow the best of geometry and of algebra, and correct all the faults of the one by the other."[1]

The strategy for mapping out the plane in Cartesian coordinates is taught in high school math. We must first choose a "home" point, an *origin*; call this point zero. We then establish a grid on the plane. First we draw the horizontal line, which we call the *x-axis*, and mark this as a number line, with zero smack in the middle. If we think of the Euclidean plane as a map, the x-axis will help establish the "longitudes" of the points on the plane. We then draw a vertical line, the *y-axis*, which intersects the x-axis at 0, and we mark this as a number line also (only, of course, this is done vertically), in such a way that the 0 of the y-axis coincides with the 0 of the x-axis, and this point we call the *origin*. So at the top of page 80 we have the Euclidean plane with its Cartesian coordinates, which we will also call the *Cartesian plane*.

Now we are ready to pinpoint any point, P, in the plane by specifying two numbers, the (Cartesian) coordinates (x,y) of the point P. The first coordinate, x, of P is the "longitude" of P. That is, P lies over the point x on the x-axis (the vertical line through P intersects the horizontal axis at x). The second coordinate of P, y, is the "latitude" of P; that is, P lies on the unique horizontal line that passes through the point (or "number") y, on the y-axis.

The unique point, for example, that is 3 units to the east of the origin and 5 units north of it has Cartesian coordinates (3,5). The unique point that is 4 units

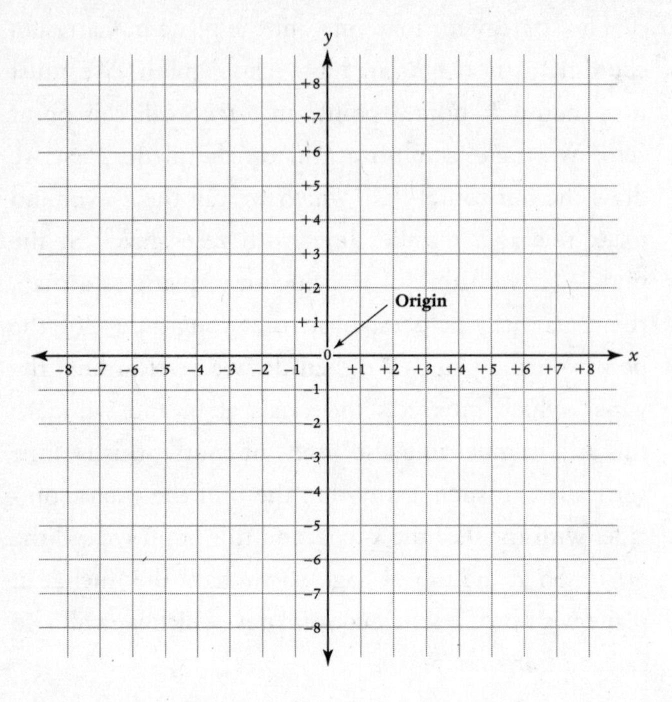

south of the origin and 2 units west of it has Cartesian coordinates (–2,–4). Remember that the first number in the parenthesis, the first *coordinate*, refers to where the point is positioned over the (horizontal) x-axis, the east-west axis; and the second *coordinate* refers to where it is positioned relative to the (vertical) y-axis, the north-south axis.

Here is the Cartesian plane showing the location of points (3,5) and (–2,–4).

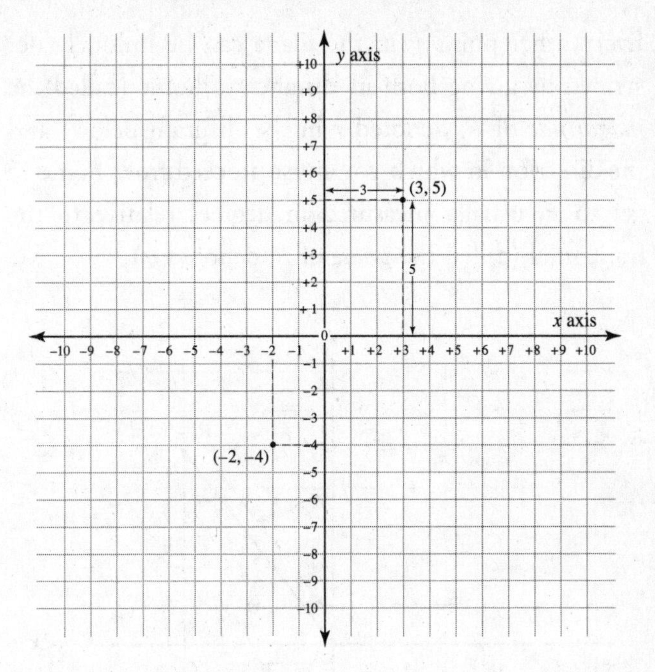

So, given any pair of numbers *a* and *b*, there is one (and only one) point *P* on the plane that carries the label—that is, has Cartesian coordinates—(*a*,*b*). An economy of expression.

The radar operator's approach uses what are usually called *polar coordinates*, and for good reason: if you are a radar operator at either the North or South Pole, your distance and direction data convert readily to latitude and longitude.

Specifically, to describe the Euclidean plane by polar coordinates you must, again, choose a starting point in the plane (home; the origin), which we again call zero.

Every other point P in the plane can be uniquely described by giving both its distance to home (called the *magnitude* of P, denoted r in the diagram below) and the direction in which you must proceed from home to get to P, usually measured in degrees relative to the horizontal (called the *phase* of P, denoted α).

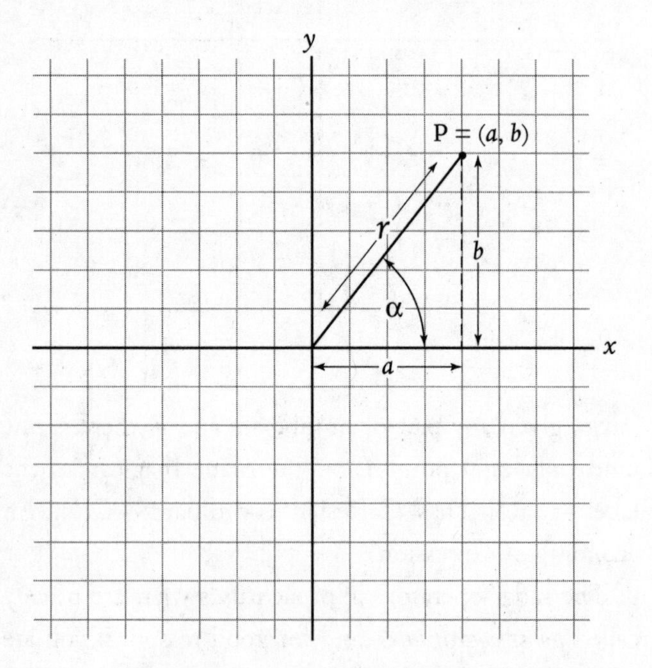

Here, P is described by its Cartesian coordinates (a,b) and by its polar coordinates: magnitude, r, and phase, α.

We have, then, two ways of naming the points on the Euclidean plane. By charting the plane as we have done, either by Cartesian coordinates or by polar coordinates, we have broken the intrinsic symmetry of points on the Euclidean plane. We have imposed upon it a specific frame of reference. We can "call up" a particular point P on the plane just by giving its coordinates, Cartesian or polar. Moreover we have a chosen "home" point, an origin to which all other points refer. And since we have *two* languages (Cartesian and polar) with which to describe the same geometric point P, we will eventually need a mode of translating one to the other.

The classical Pythagorean theorem (applied to the triangle in our diagram) helps to provide us with a clean algebraic formula for the *magnitude*, r, of P if we know its Cartesian coordinates (a,b):

$$r = \sqrt{a^2 + b^2}.$$

But to express the *phase*, α, of P (assuming that P is not the origin) in terms of its Cartesian coordinates, or to get a point's Cartesian coordinates from its magnitude and phase, is not as simple. This task, you might remember from high school, is one of the missions of the subject of *trigonometry*: the ratios a/r and b/r depend only on the phase α of P and are called the *cosine* and the *sine* (respectively) of the angle α. So if you happen to know $\cos(\alpha)$ and $\sin(\alpha)$, you can retrieve the Cartesian coordinates (a,b) of P from them and the magnitude r:

$$a = r.\cos(\alpha) \text{ and } b = r.\sin(\alpha).$$

If you only know cos(α) and sin(α) approximately, you can, at least, approximate the Cartesian coordinates of *P*.

24. The geometry of qualities.

There are striking precursors to these modes of charting Euclidean geometry. Let us read a bit from the fourteenth-century treatise written by Nicolas of Oresme, *Tractatus de Configurationibis Qualitatum et Motuum* [A Treatise on the Configuration of Qualities and Motions].[2] Before we do so, I should prepare you for Oresme's vocabulary. Oresme, in his treatise, is explaining his discovery of what we, in modern times, would call *drawing graphs.* What is a graph? Say you have a range of items (the years 1950, 1960, 1970, for example), and for each of these items you want to make a measurement, a "reading" of something (the population of the United States, for example). Oresme would call the "range of items," whatever they are (e.g., the years), the *subject.* He would call the thing that is to be measured for each item (e.g., the U.S. population) the *quality*.

How do we ordinarily make a graph? We first lay out, horizontally usually (on the *x*-axis, we might say), the range of items. And in some pictorial way, such as by erecting a vertical bar over each item, or plotting a point at some height over each item, we *exhibit* the data; the length of the bar, or the height of the point, is an indicator of the size, as we have measured it, for each of the items. Nowadays we tend to require that our measurements be actual numerical quantities, but for Oresme, only the *proportions* of measurements, a comparison of the measurement made for one item to that made for another, are relevant. These proportions Oresme refers to as *intensities (of the quality)*, and his great aim is to picture the variation of intensities, as one ranges from item to item. In his vocabulary, he wanted to *see the configuration of qualities.*

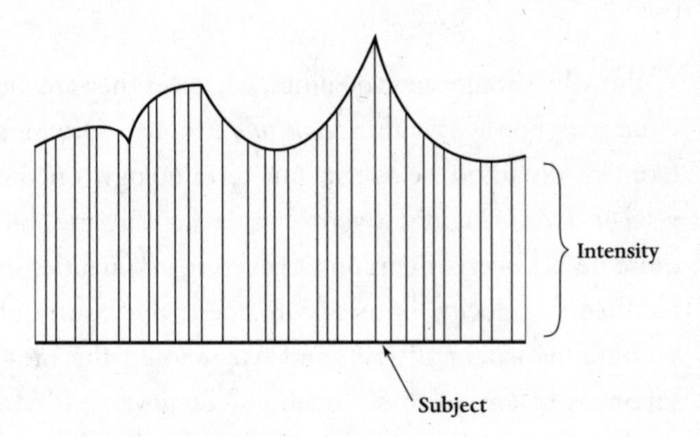

Oresme opens his treatise with a discussion of the nature of the readings themselves, in a paragraph entitled "On the Continuity of Intensity":

> Every measurable thing except numbers is imagined in the manner of continuous quantity. Therefore, for the mensuration of such a thing, it is necessary that points, lines, and surfaces, or their properties, be imagined. For in them (i.e., the geometrical entities), as the Philosopher has it, measure or ratio is initially found, while in other things it is recognized by similarity as they are being referred by the intellect to them (i.e., to geometrical entities). Although indivisible points, or lines, are nonexistent, still it is necessary to feign them mathematically for the measures of things and for the understanding of their ratios. Therefore every intensity . . . ought to be imagined by a straight line perpendicularly erected on some point of the space or subject of the intensible thing, e.g., a quality.

But why should these qualities, whatever they are, be visualized linearly, within *continua* (to use Oresme's term, as discussed below; in our terminology, on the *y*-axis)? This is hardly a simple issue for Oresme, because he has every intention of covering *all* qualities in his theory (velocity, hotness, whiteness, the beauty of sounds, the extent of joy); and why should the measurement of any of these "qualities" be governed by a single *number* (as we would say) or a single *length* (as he would say)? This question is reminiscent of the discussions about IQ and how unlikely it is that a single number is sufficient to evaluate human intelligence.[3] The corresponding issue, about the measurement of the *intensity* of his *qualities*, bothered Oresme, and he returns to it, giving the following curious argument:

> Again, intensity is that according to which something is said to be "more such and such," and "more white," or "more swift." Since intensity, or rather the intensity of a point, is infinitely divisible in the manner of a continuum in only one way, therefore there is no more fitting way for it to be imagined than by that species of a continuum which is initially divisible and only in one way, namely by a line. And since the quantity or ratio of lines is better known and is more readily conceived by us . . . therefore such intensity ought to be imagined by lines and most fittingly by lines which are erected perpendicular to the subject.

Oresme's extraordinary argument has two parts, each introduced with the word *since*. First, "intensities," Oresme claims, are "infinitely divisible . . . in only one way." Moreover, "a continuum" (meaning, I suppose, a straight line segment, a *y*-axis in our vocabulary) has these same characteristics: it is "infinitely divisible and only in one way." So Oresme's first argument is that salient characteristics of "intensities" and of "continua" match perfectly. One might imagine he would consider the issue sewn up at this point. But he goes on with a second argument.

Oresme prefers to line up his "intensities" in a continuum, "since," he says, continua are "better known and . . . more readily conceived by us." He concludes that "therefore" the intensities "ought to be imagined" in this way. The logic here is that our mere familiarity with a tool that is conceivably of use for the imagination is enough reason for us to make use of it, rather than to look elsewhere for other tools. According to Oresme's argument, we ought to make the most of what we are most familiar with: there is virtue, he seems to claim, in practicing an economy of the imagination.

25. The spareness of the inventory of the imagination.

Jorge Luis Borges, while analyzing in one of his lectures the phrase "the stars looked down," claimed that in contrast to what is usually assumed about poetry—that poetic tradition draws from virtually unlimited sources for its metaphorical material—it is, rather, the reverse that is true.[4] Only a very restricted collection of images is traditionally called upon to supply metaphors for poetry. Moreover, this fact helps to increase the intensity of any further metaphoric use of this rare, shared material.

Stars, for example, are everywhere in poems. Nevertheless, the emotional tone of the phrase "the stars looked down," taken in isolation, is as yet indeterminate. Those stars could present a benevolence blanketing us with their warmth, they could be comforting witnesses that all's right with the cosmos, or else they could be glaring down, affecting a sinister penetration, a cold perception of our unimportant place in that cosmos. The poet, using such a phrase, has the liberty of carving an exquisite, meticulously specific, emotional context for the phrase, and to achieve this, can call up the full resonance of the metaphoric tradition regarding the image of stars.

Let us briefly entertain this Borgesian conceit that the greenhouse of poetic tradition grows but few species of metaphor, and let us ask whether this may affect our reading of the phrase "the yellow of the tulip."

How prevalent are tulips in poems and what do they signify? They play their passionate role in the poetry of Omar Khayyám and of Hāfez. They are a sign of God's mystical joy in the Rumi poem "Unmarked Boxes":

> *Part of the self leaves the body when we sleep*
> *and changes place. You might say "Last night*
> *I was a cypress tree, a small bed of tulips,*
> *a field of grapevines." . . .*[5]

But in English poetry, tulips seem to "bloom as they are told." They have the staying power of a watercress sandwich:

> *You are a tulip seen to-day,*
> *But, dearest, of so short a stay.*[6]

Tiptoeing through any edition of the *Oxford Book of English Verse*, you can gather bouquets of roses, daffodils, and unnamed weeds, but few tulips. The lily of the field can fester, but not, at least by repute, the tulip. Could it be that the combination of the ubiquity of floral poetic imagery, together with the relatively unsung image of the tulip, makes the tulip, for us, *more* potent in our imagination? And it is indeed potent, as in Chase Twichell's poem "Tulip," whose (necessarily) terminal stanza reads:

> *Look, a yellow tulip*
> *in the charcoal sky—*
> *a vividness passing so quickly*
> *I have to abandon the poem*
> *to follow it.*[7]

6

JUSTIFYING LAWS

26. "Laws" and why we believe them.

We were left in section 22 with the question of whether to believe the distributive law,

$$A \times C + B \times C = (A + B) \times C,$$

holds for all real numbers, positive or negative, and if we believe so, why we do. This equation, translated from symbols to words, says that to multiply a given number C by the sum of two numbers is the same as multiplying C by each of those numbers, and then adding the results. The distributive law relates *two* operations, one to another: addition and multiplication. Perhaps, then, if we are to get practice in justifying laws, we should first try our hands on simpler laws. For example, laws involving only a single operation.

Here is one: we all know that 3 + 5 is the same as

5 + 3. Or, more generally, in *adding* two numbers it does not matter which comes first. In symbols:

$$A + B = B + A.$$

And here is another law:

$$A \times B = B \times A.$$

That is, in multiplying two numbers it is irrelevant which of them comes first. For example,

$$5 \times 8 = 8 \times 5.$$

How can we justify, say, the second of these laws:

$$A \times B = B \times A?$$

Before we get started, we must be clear how "generally" we wish to make our justification. That is, which A's and B's are we thinking of when we claim the truth of the equation $A \times B = B \times A$? Suppose we want to prove this equation holds where A and B are allowed to be (any) positive whole numbers. As an exercise, determine whether you think the following series of pictures persuades you of its truth.[1]

In words,

5 lines of arrows with 8 arrows in each line
equals
8 lines of arrows with 5 arrows in each line

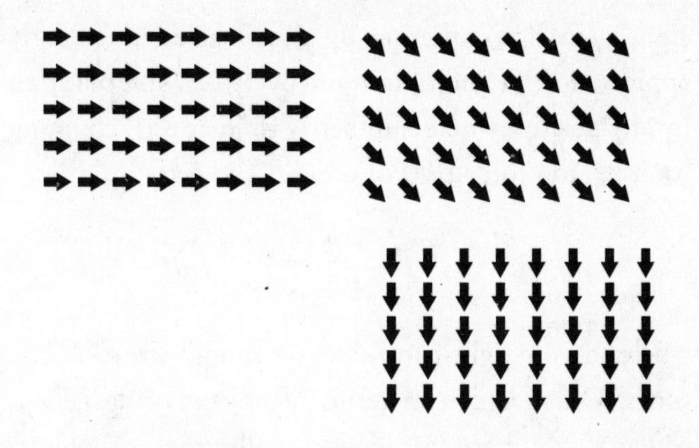

I find these pictures to hold a very compelling explanation of why $A \times B = B \times A$ for all positive numbers A and B. But surely one cannot be fully persuaded by the pictures until one extracts the essential argument implicit in them.

If these pictures persuade you that $A \times B = B \times A$ is right, it is a good exercise to try to say why they do. But if you wish to do this exercise, before you get started you must have a clear idea of what it means to *multiply*. As Aristotle took great pains to explain to us, in order to prove anything, one must first define one's terms. Have we pushed our issues back yet another notch? Do you smell a whiff of infinite regression in the air? Don't lose heart.

27. Defining the operation of multiplication.

Spend any time with numbers and you encounter multiplication. The chant that begins "Two, four, six, eight"—that is, counting by twos—gives us an early appreciation of multiplication by two. Multiplication by any positive whole number N is, in effect, "counting by N's." You tote up M of these N's to get

$$M \times N = \underbrace{N + N + \ldots + N}_{M},$$

where, on the right-hand side, we should write "N" M times. Or, to put it in terms of objects being tallied, suppose you have M distinct collections of objects (M shelves of CD's, say), any two of these collections having the same number, N, of objects (N CD's per shelf). Then $M \times N$ is the total number of objects (CD's) in the conglomerate made by combining all of your collections.

We are counting by N's all the time. When faced with the chore of counting a large number of things, we often structure the activity by putting the things in equal-size piles, and then counting the piles. This gentle organization of our task helps to avoid the sort of confusion faced by Alice (in *Through the Looking Glass*), when she is challenged by the White Queen with the arithmetic question of tallying "one and one and one and one and one and one and one and one and one and one."

Even if we just count a straight 1, 2, 3, . . . without imposing modular stopping points in our calculation, as soon as we get high enough the very language for numbers (e.g., one hundred, one hundred and one, one hundred and two, . . .) imposes such a structure for us. But we also count by an odd assortment of *N*'s (dozens, scores). Or think of Herman Melville's description, in *Moby Dick*, of the Polynesian harpooneer Queequeg in the Spouter-Inn, counting the pages, in fifties, of a certain large "marvelous book." At the end of each fiftieth page, Queequeg would give

> utterance to a long-drawn gurgling whistle of astonishment. He would then begin the next fifty; seeming to commence at the number one each time, as though he could not count more than fifty, and it was only by such a large number of fifties being found together, that his astonishment at the multitude of pages was excited.[2]*

Give a computer any pair of numbers, *M* and *N*, of under, say, a hundred digits, and the computer can multiply them in nanoseconds. The job of performing the operation of multiplication is therefore more rapidly done than the job of defining this operation. Our earlier description of the operation of multiplication of positive whole numbers in the context of "collections of objects" can be refashioned to provide a perfectly formal definition of multiplication ("of sets"). Mathe-

*Melville intends, I imagine, Queequeg's self-constructed *numeracy* as a backdrop for his self-constructed painstaking *literacy*, shown later in the signing of his mark.

maticians also make use of two other approaches to formally defining the operation of multiplication (for positive whole numbers).

The first of these provides a *definition*, which also might substitute as a method (although a very slow method!) for multiplying numbers. I will call this the "creeping strategy."[3] The second approach is to give a *structural characterization* of the operation of multiplication: among all conceivable "operations" that can be performed where the input is a pair of positive whole numbers, how can you characterize, by simple laws, the operation that multiplies them together?

The *creeping strategy* builds the operation of multiplication of positive whole numbers up from its beginnings, from the operation of addition. The strategy plays on the fact that 1 times any number N is just N; 2 times N is $N + N$. And if for a particular number M you know what M times N is, you have this simple formula telling you what $M + 1$ times N is (if you can add):

$$(M + 1) \times N = M \times N + N \tag{6.1}$$

So, creeping up, starting with $M = 1$ and successively applying equation (6.1), you can work out what any positive whole number times N is.

Do you see a bit of the distributive law peeking through equation (6.1)?

Here, more explicitly, is how you creep up. We know

how to multiply 45 by 1, so applying equation (6.1), with $M = 1$, $N = 45$:

$$2 \times 45 = 1 \times 45 + 45 = 90.$$

So now we know what 2×45 is. Applying equation (6.1) again, with $M = 2$, $N = 45$, gives

$$3 \times 45 = 2 \times 45 + 45 = 90 + 45 = 135.$$

Now we know what 3×45 is. Successive applications of equation (6.1) will tell you how to multiply 45 by anything. For example, assuming that you have worked your way up to $M = 12$, and know that $12 \times 45 = 540$, an application of our equation will tell you that

$$13 \times 45 = 12 \times 45 + 45 = 540 + 45 = 585.$$

The *structural characterization*, as I mentioned, singles out the operation of multiplication (of positive whole numbers) from among all operations. Suppose that you have some "unknown" operation that allows, as its "input," any two positive whole numbers M and N and produces, as a result, another whole number, which we will denote $M * N$. So, if you are given any two positive whole numbers, say 13 and 45, you can perform this "mystery operation" on these two numbers to get a result (which we would denote $13 * 45$). Suppose further that we know our mystery operation satisfies the two simple laws (*a*) and (*b*):

(*a*) For any positive whole number N,

$$1 * N = N.$$

(*b*) For any three positive whole numbers A, B, and C,

$$A * C + B * C = (A + B) * C;$$

that is, the *distributive law* holds.

Then we can show that our mystery operation $*$ is none other than multiplication. That is,

$$13 * 45 = 13 \times 45 = 585,$$

and, more generally, $M * N = M \times N$ for any two positive whole numbers M and N.

The distributive law *characterizes* multiplication of positive whole numbers. It is the engine of multiplication.[4]

A formal treatment of our subject would offer, at this point, a proof that these laws—(*a*) and (*b*)—do indeed characterize the operation of multiplication, as defined, say, by the creeping strategy. Instead, let us limit ourselves to asking: Does the following diagram convince you that the distributive law holds for positive whole numbers?

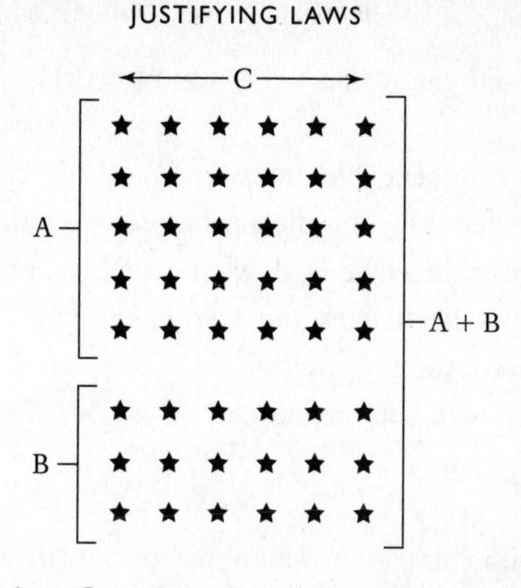

$$A \times C + B \times C = (A + B) \times C$$

28. The distributive law and its momentum.

To think meaningfully about "minus times minus" you must have a definition of the operation of multiplication that applies to negative numbers. To be fully inclusive, of course, we should ask about "minus times plus" and "plus times minus" as well.

Now, nothing stops you from proclaiming, by fiat, the "law"

$$(-M) \times (-N) = M \times N,$$

for any two positive whole numbers M and N. If you do, your answer to Stendhal's query—"Why is minus times minus equal to plus?"—would be that it is so *by definition.* You have willed it to be so. But Stendhal would then surely ask you this follow-up question:

"Why did you define it this way, rather than any other way?"

I recommend this answer: You have chosen the (only) definition of the operation of multiplication, valid over the range of all whole numbers, positive and negative, that retains and extends the basic structural characterization of multiplication of positive numbers. That is, with your definition,

$$(-M) \times (-N) = M \times N,$$

and with companion definitions to cover the cases "minus times plus" and "plus times minus,"*

$$(-M) \times N = -(M \times N) \quad \text{and} \quad M \times (-N) = -(M \times N),$$

the distributive law,

$$(A + B) \times C = A \times C + B \times C,$$

holds for all whole numbers, and your definitions provide the *only* extension of the operation of multiplication that satisfies this distributive law.

Stendhal, however, might then stubbornly ask: "Why should we assume the validity of the distributive law over this extended range?" You might worry, at this point, that no matter how you try to justify the distributive law in its extended range—over positive and negative numbers—you will, of necessity, have to fall back

*Stendhal does not seem to regard these cases as problematic. Do you?

on some definition of multiplication for this extended range.

29. Virtuous circles versus vicious circles.

Are we going around in circles? We have tried to answer Stendhal's query "Why is minus times minus equal to plus?" And we justified this on the basis of the distributive law. We then asked ourselves how we can justify the distributive law itself. Any justification of it must depend upon our definition of the operation of multiplication, including our definition of that operation when applied to negative numbers—which is where we began.

Does coming around full circle mean that we have done nothing? I think not. "Circles" in trains of thought have a bad press. "That definition is circular!" is a complaint often meant to demolish the worth of the definition in question. Yet many of our most important arguments and definitions are unavoidably circular, and their depth derives precisely from the tightness of the circle they draw. Algebra, for example, could not get very far if it avoided objects defined "circularly" (such as "that number equal to its own square minus 6").

The distributive law and our definition of multiplication over the full range of numbers, positive and negative, fit perfectly together, one justifying the other, the other justifying the one.

30. So, why does minus times minus equal plus?

Here is a summary of what we have achieved so far. The initial stumbling block regarding "minus times minus" was that although we all have a clear understanding of what it means to multiply a positive whole number by a positive whole number, the question "Does minus times minus equal plus?" cannot be tackled until we know what it means to multiply a negative number by a negative number. Utterly equally, the question "Does minus times plus equal minus?" cannot be answered until we have a working definition of multiplication of positive by negative numbers. This is only fair, because without even a working definition, we can't work. We had to go back to the drawing board to provide these definitions.

Back at the drawing board, we explicitly defined, or characterized, the operation of multiplication in the cases where we thought we knew what we were doing (i.e., for positive whole numbers). We came to the conclusion that the *distributive law* is a fundamental characterization of the operation of multiplication (of positive whole numbers).

We wanted, then, to maintain this "fundamental characterization" as we extended the range of the operation of multiplication from positive whole numbers to all whole numbers. Happily (although we haven't proved this here), there is, in fact, *only one way* to extend the definition of multiplication to all whole numbers, negative as well as positive, if we wish (we do!) 1 times any number N to equal N, and if we wish (we do!) the distributive law to hold. We then adopt this unique (extended) definition of multiplication.

Given this (extended) definition, we have seen that minus times minus is plus.

PART II

7

BOMBELLI'S PUZZLE

31. The argument between Cardano and Tartaglia.

Violent argument surrounded the method of solving one of the mathematical problems described by Girolamo Cardano in his *Ars Magna.* In contrast to the usual form taken by priority disputes in science, this dispute was not, at least at first, about who originated this method. Cardano acknowledges (three times, in different parts of his book) that he learned this method from his friend Niccolò Tartaglia, and moreover, Cardano writes,

> Scipio Ferro* of Bologna well-nigh thirty years ago discovered this rule and handed it on to Antonio

*Scipione de Floriano de Geri dal Ferro (c. 1465–1526) was also referred to as Dal Ferro, Del Ferro, Ferro, and Ferreo in writings of the time. I will call him Dal Ferro, since this is the name his contemporaries most often used and is the name used by Bortolotti, who rediscovered Dal Ferro's original papers in the library of the University of Bologna, and who is one of the editors of Bombelli.

> Maria Fiore of Venice, whose contest with Niccolò Tartaglia of Brescia gave Niccolò occasion to discover it. He [Tartaglia] gave it to me in response to my entreaties, though withholding the demonstration.[1]

Reading this, you might be surprised to see the word *contest*. What contest?

In 1535, upon returning to Venice after studying in Bologna, Fiore challenged Tartaglia to a *public* problem-solving contest. I find it hard to reconstruct the setting here. Was this a kind of intellectual *palio*?* Such problem duels were common: the challenger would make the problem contest known through a public poster or a proclamation, called the *cartello*, in which a challenge (the *sfido*) would be thrown out to a specific rival. The ultimate affair was referred to as the duel (*duello*). I assume it would have been accompanied with some fanfare, with "seconds," with food and drink. How many were in the audience, and where did they sit?

Fiore set the problems in the public contest, and he did this with a distinct advantage, for he had at his disposal the powerful method of his teacher Dal Ferro. The problems he posed were, in effect, to find the solutions of cubic equations. Here is a sample collection of problems, with some commentary. Find all solutions of

*The *palio* is a traditional horse race held each summer in Siena, in the Piazza del Campo—a magnificent space, but not one a dispassionate observer might judge really suitable for horse racing.

$$X^3 = 6X + 40,$$

or

$$X^3 = X + 1,$$

or

$$X^3 + 1 = 3X.$$

As we have already seen, there is some ambiguity in the phrase "find all solutions." Do we mean all solutions that are positive whole numbers, as was the case with Bháskara's problem about bees (see sect. 4)? Do we mean all whole-number solutions, whether positive or negative? Do we allow real-number solutions (recall the discussion of real numbers in section 19)? Do we discard as "impossible solutions" expressions that involve imaginary numbers but that nevertheless seem to represent "formal solutions" to our equations, as Chuquet did with his candidate solutions $3/2 + \sqrt{-1.75}$ and $3/2 - \sqrt{-1.75}$ (see sect. 4)?

The public-contest problems would not, of course, be set in the algebraic language displayed above, but would be given, rather, as word problems. The first problem, $X^3 = 6X + 40$ (find the quantity that has the property that its cube is equal to 6 times itself plus 40), would probably not have been posed as a challenge, because its answers would be too easily guessed by seasoned contestants. The cube of the integer 4 (i.e., 64) is

6 times 4 plus 40. That is, $X = 4$ is a solution of the equation $X^3 = 6X + 40$. It is, in fact, the only integer solution of this equation.[2]

The second problem, $X^3 = X + 1$, has a unique real-number solution,[3] and, I imagine, would be fair game as a challenge question, since it is amenable to the method that was, if not in the air, at least in the secret repertoire of some of the lucky contestants. I will (very briefly) discuss this method later.

The third problem, however, $X^3 + 1 = 3X$, has three different (real-number) solutions and *would have seemed* out of reach of the method that works for the second problem. I imagine this third problem would not have been fair game as a challenge question. The question of how to find the solutions of cubic equations that have three distinct real solutions, if none of these solutions can be easily guessed, later induced Rafael Bombelli to press the "standard method" to the extreme. It also led him into somewhat uncharted territory and presented a puzzle that occupied him throughout the twenty years or so that he spent writing his treatise *L'Algebra*, as we shall see later in this chapter.

But these problems seem hardly fit for public gladiator displays. The setting here reminds me of nothing so much as the Monty Python skit depicting Thomas Hardy composing his *Return of the Native* in agonized cogitation, in a football stadium before a wildly cheering sports crowd, and the "letter-by-letter" description, so to speak, being reported by a quite keyed-up sportscaster.

Tartaglia was decidedly the underdog in the problem contest, lacking the powerful method that Fiore had. In the written record Tartaglia left, he describes the agonies he suffered before the contest.[4] For it wasn't until the very night before that Tartaglia rediscovered the precious method his rival had learned from Dal Ferro—but as a result, says Oystein Ore in his foreword to *Ars Magna*, "Fiore suffered a humiliating defeat."[5]

After the contest, Cardano repeatedly requested that Tartaglia explain his method to him. Tartaglia, in a meeting with Cardano of which we have contradictory written accounts by the two interlocutors, finally gave in. The later feud between Tartaglia and Cardano focused not on who discovered the method but on whether in this meeting Tartaglia gave Cardano permission to publish his method (irrespective of whether or not he gave due credit to Tartaglia). Here is Ore's account:

> [A]ccording to Tartaglia, Cardano swore a most solemn oath, by the Sacred Gospels and his word as a gentleman, never to publish the method, and he pledged by his Christian faith to put it down in cipher, so that it would be unintelligible to anyone after his death.[6]

This description of the events was contradicted by Lodovico Ferrari (a student of Cardano, and his secretary; he later produced methods for solving equations of the fourth degree), who claimed to be present at this meeting. Ferrari said that the issue of secrecy was never discussed there, and in a later challenge to Tartaglia (to meet him in a public dispute on scientific questions for a prize of up to 200 *scudi*), Ferrari threw down his gauntlet:

> This I have proposed to make known, that you have written things which falsely and unworthily slander the above-mentioned Signor Gerolamo [Cardano] compared to whom you are hardly worth mentioning.[7]

There is a striking difference in diction between these taunts, steeped in the amour propre of the late Renaissance, and the tone of Bombelli's treatise, published at roughly the same time.

32. Bombelli's *L'Algebra*.

Rafael Bombelli (1526–72) wrote his text *L'Algebra* over a span of two decades in the middle of the sixteenth century. The title of his book, which is also the name of his subject, reflects the enormous influence of Arabic sources, notably the ninth-century work *Kitāb al-jabr wa al-muqābalah* of Muhammad ibn Mūsā al-Khwārizmī. Bombelli himself, in his judgment of al-

Khwārizmī's treatise,[8] may not have been sufficiently sensitive to its originality, but at least one modern scholar perceives in al-Khwārizmī's writings a fresh, unified view (geometric and algebraic) of the role of the "unknown,"[9] a view unappreciated by his predecessors or contempories, one allowing for a focused concentration on the mechanics of algebra. The *al-jabr* and *al-muqābalah* of the Arabic title,[10] while referring to the (new) subject as a whole, also refer to two specific processes. *Al-jabr* is the operation of moving quantities from one side of an equation to the other, changing sign; *al-muqābalah* is the operation of collecting "like" terms.

Bombelli was a civil engineer involved in the project of draining the Chiana swamp in Tuscany, and only during periods of interruption of this project was he actively engaged in writing. His text *appears* to have an orderly, paced, rather modern architecture. At first glance, it would seem more like a systematic, theory-driven work such as the college texts we are familiar with nowadays than like any of the treatments of mathematics composed by Bombelli's contemporaries. It lacks the trumpet calls of Cardano's *Ars Magna*, published just a few years earlier, or of Viète's flamboyant memoir that we discussed earlier (sect. 20). On the other hand, *L'Algebra* also reads, in spots, as if it were a private journal: it records Bombelli's hesitations, puzzlements, and changing attitudes over the twenty-year period. This is a happy accident for us, for we will be considering one of these changes of viewpoint later: the way in which Bombelli confronts certain "cubic radicals." Bombelli wrote in Italian (*L'Algebra* is possibly the first long treatise on mathematics written in Italian), and he was therefore obliged to invent, as Dante did, an Italian vocabulary for his subject.

Bombelli's work is divided into five volumes: the first cleanly deals with basic numerical operations (squaring, cubing, taking square roots, taking cube roots, etc.); the second provides a theoretical underpinning for algebra per se, dealing with the notion of an unknown and solutions of cubic equations; and the third is organized as a series of problems: the applications, as it were, of Bombelli's theory. Only these first three volumes were published in Bombelli's lifetime. Bombelli apologized for this, saying that he could not publish the rest because it had not yet been "brought to the level of perfection required by mathematics."

In contrast to the worldly, or sensual, word-problems that color Hindu and early Italian texts, Bombelli's problems have a purer format. He boasts:

> [I have] almost totally deviated from the methods employed by the writers of this discipline, who for the most part, when treating mathematical problems always cloak them beneath the veil of such activities and business as men normally conduct (like selling, buying, repaying and exchanging money; amassing

> interest; taking deductions; valuing money, metals, and weights; [calculating] the losses and gains of a company; gaming; and an infinite number of other human activities with which the aforesaid books are crammed full, as you can see), while I alone have left the problems to the dignity of arithmetic . . .[11]

The first essential appearance in *L'Algebra* of numbers involving square roots of negative quantities seems to be a somewhat accidental intrusion. They occur in the midst of an algebraic formula, and they seem to take Bombelli by surprise. "I have found a new type of *cube* root which behaves very differently from the others,"* he muses, and then goes on to say that these cube roots "will seem to many merely sophistic and not real, which was the opinion even I held."

These *quantities* (if so they are) have stumbled into Bombelli's text, and hardly in a simple form. They enter the scene as participants in a formula that is a sum of cube roots of the type of numbers that had given Cardano mental tortures. But the tongue-twisting complication of the way in which they first present themselves is not unusual: it is rare for any new concept to have a clean "theorified" look in its debut appearance, the look it eventually will be given.

*Yes, "cube root," for Bombelli's imaginary numbers make their entrance as parts of the general algorithm for solving cubic equations. See page 133 of Bombelli's *L'Algebra*.

What kind of mathematical objects are Bombelli's "cube roots"? What does he think they are? Although Bombelli pays great homage to Cardano elsewhere in his book, he does not mention Cardano or anyone else at this point. It is natural to assume that the reason he neglects to mention his predecessors is that he simply doesn't know whether his "new types of cube roots" are, in fact, *always* also expressible in terms of the imaginary numbers, the "sophistic negatives," of Cardano. For Bombelli, his new types of roots are, in general, truly *new*, if they exist! And he does come to believe they do exist, but he never exactly says what he means by "exist." To read Bombelli's treatise is to follow Bombelli's long involvement with the questions: *Do these cubic radicals exist? What do they mean?*

It is time to explain what these roots are and why Bombelli wants them.

33. "I have found another kind of cubic radical which is very different from the others."

Earlier we discussed the "general" formula that solves quadratic equations (see sect. 8). If you are given a quadratic equation

$$X^2 + bX + c = 0,$$

where b and c are particular numbers and you are seeking a number X that satisfies this equation, then there are (generally) two values of X that "work." Here they are again—in the *quadratic formula*:

$$X = \frac{-b + \sqrt{b^2 - 4c}}{2} \text{ and } X = \frac{-b - \sqrt{b^2 - 4c}}{2}$$

Bombelli was considering the similar problem for cubic polynomials, and more particularly for polynomial equations of the form[12]

$$X^3 = bX + c,$$

where we think of b and c as specific numbers that have been given to us and we are looking for values of X that satisfy this equation. Bombelli was working in the tradition of Cardano, Tartaglia, and Dal Ferro, and therefore had recourse to a general expression, Dal Ferro's formula, that in many cases gives those values, at least in terms of square roots and cube roots. Dal Ferro's formula is, perhaps, a bit more complicated than its counterpart that solves quadratic equations, but not much more. We will be parsing it slowly, so don't be too dismayed by its shape. Here it is:

$$X = \sqrt[3]{\frac{c}{2} + \sqrt{\frac{c^2}{4} - \frac{b^3}{27}}} + \sqrt[3]{\frac{c}{2} - \sqrt{\frac{c^2}{4} - \frac{b^3}{27}}},$$

Dal Ferro's formula

If you are puzzled by this expression, you should be, and you are in good sixteenth- and seventeenth-century company. As in the case of the quadratic formula, if you are courageous, and adept at cubing things, you can try your hand at cubing the expression for X above, and you will find that it (whatever *it* is) is equal to $bX + c$. That is, Dal Ferro's formula "works," whatever the formula means. But if this were the end of the story, it would be of limited use. One of the fruits of our work in imagining imaginary numbers will be (in chap. 12) that we will be able to give a clear interpretation of Dal Ferro's formula that makes the formula do its job (solve cubic equations) exquisitely well.

Sometimes Dal Ferro's formula does indeed give us, without too much work on our part, an honest solution to the equation $X^3 = bX + c$. Let us examine more closely the monster formula.

When the quantity $c^2/4 - b^3/27$ is a positive real number, we can take its square root, as we are requested to do by Dal Ferro's formula, and find that the two expressions nestled under the two cube root signs ($\sqrt[3]{\ }$) are real numbers. And since any real number has a unique (real-number) cube root, we can evaluate the formula to find a real-number value for X, and this will indeed solve the equation $X^3 = bX + c$. The manner in which our formula is written tells us how to compute it: find the square root of $c^2/4 - b^3/27$ and alternately add this to and subtract this from the quantity $c/2$ to get two real numbers, then find the cube roots of each of these numbers and sum them up. Luckily, in this case (i.e., when $c^2/4 - b^3/27$ is positive), it turns out that $X^3 = bX + c$ has *only one* real-number solution. You have therefore solved the problem!

Solved the problem? Well, here is a more accurate assessment of what Dal Ferro's formula gives us. The formula does the analogous job for cubic equations that the quadratic formula does for quadratic equations. If you want to solve the "general" cubic equation $X^3 = bX + c$, all you must be able to do is extract square roots and cube roots and perform standard algebraic operations with them. The formula, then, has reduced the problem of solving general cubic equations to the problem of extracting square and cube roots. All this, when $c^2/4 - b^3/27$ is positive, was well understood by Bombelli's predecessors.

Bombelli, however, was interested in pursuing the opposite, more obstreperous case—that is, when $c^2/4 - b^3/27$ is a negative real number. Let us refer to this number $d = c^2/4 - b^3/27$ as the *indicator* of our cubic equation.[13] Our equation behaves quite differently depending upon whether the indicator $d = c^2/4 - b^3/27$ is positive, negative, or zero. When the indicator is positive (the easy case we have already discussed), $X^3 = bX + c$ has a unique real-number solution, and this solution is given by Dal Ferro's formula. But when the indicator is negative (the case that caught Bombelli's attention), one is faced with something of a puzzle.

First (as it will turn out), $X^3 = bX + c$ has *three* distinct real-number solutions. We have, then, three numbers to find. Second, Dal Ferro's formula,

$$X = \sqrt[3]{\frac{c}{2} + \sqrt{\frac{c^2}{4} - \frac{b^3}{27}}} + \sqrt[3]{\frac{c}{2} - \sqrt{\frac{c^2}{4} - \frac{b^3}{27}}},$$

which we can abbreviate, using our notation for the indicator $d = c^2/4 - b^3/27$, to read

$$X = \sqrt[3]{\frac{c}{2} + \sqrt{d}} + \sqrt[3]{\frac{c}{2} - \sqrt{d}},$$

now requires taking the cube roots of numbers that are themselves not real numbers. Do these cube roots *exist*? In what sense do they *exist*? In any event, the formula hardly seems an adequate guide to finding the three real-number solutions to our cubic equation $X^3 = bX + c$. More puzzling still is that Dal Ferro's expression, the right-hand side of Dal Ferro's formula, satisfies—in a naive sense—the requirements of being a solution to our equation. If you replace X by that expression and just do the algebra, you discover that, whatever this X means,

$$X^3 = bX + c.$$

You might think this state of affairs is hardly any different from the situation considered earlier by Chuquet, when he sought a number whose triple is 4 plus its square and discovered that his method led to solutions that were not real numbers. Chuquet called those solutions "impossible," thereby correctly solving his problem: there is *no* real number whose triple is 4 plus its square. But Bombelli is in a stranger situation, for his problem *does* have real-number solutions, in fact, three of them. And he is contemplating a troubling expression (Dal Ferro's) that, at the same time, seems to solve his problem and yet provides none of its solutions!

To ease into Bombelli's puzzle, let us examine a situation where the indicator d ($= c^2/4 - b^3/27$) is zero. Take the equation

$$X^3 = 3X - 2;$$

that is, we are specializing the general equation by putting $b = 3$ and $c = -2$. So the indicator $c^2/4 - b^3/27$ is $2^2/4 - 3^3/27$, which is zero. This equation has precisely *two* solutions ($X = 1, -2$). Let us see how "good" Dal Ferro's formula is at "picking out" those two solutions.

Putting $b = 3$ and $c = -2$ in the formula, we calculate

$$X = \sqrt[3]{-1} + \sqrt[3]{-1}.$$

What kind of "thing" is this expression? How can we interpret it?

We are now in a funny situation vis-à-vis our problem (solve $X^3 = 3X - 2$). For we actually know the solutions ($X = 1$ and $X = -2$) but we have a curious expression (a sum of two cube roots of -1) that purports to provide these solutions. And our puzzle is to find an appropriate interpretation of this curious expression that yields those two solutions. Later (sect. 43) we will see why this curious expression gives *both solutions*, $X = 1$ and $X = -2$, to our equation $X^3 = 3X - 2$. But even now you might be able to think of a natural interpretation of Dal Ferro's expression so as to have it yield one of the two solutions (see endnote).[14]

Here is what Bombelli says about his use of Dal Ferro's expression (in cases where the indicator is not positive):

> This kind of root has in its calculation [*algorismo*] different operations than the others and has a different name . . . [It] will seem to most people more sophistic than real. This was the opinion I held, too, until I found its geometrical proof.[15]

What sort of "geometrical proof" does Bombelli have in mind? The fun is to read the fifth book of Bombelli's *L'Algebra*[16] (unpublished in his lifetime), in which there are hints that he understood that angle trisection problems lead to cubic equations and the solution of an appropriate cubic equation could help to trisect a given angle. Is this the kind of "geometric proof" that Bombelli had?

Such a connection between angle trisection problems and cubic equations is entirely explicit in the later *Analytic Art* of Viète.[17] Viète also understood the relationship between dividing angles into five equal parts and fifth-degree polynomial equations, and the same for some higher degrees. This connection between angle division and polynomials is "early trigonometry." We will take up this subject in earnest in part III, and we will be using it (in sect. 63) to find solutions to cubic equations.

Despite his reflections and initial doubts about its existence, Bombelli permits himself to work with his new type of root. Reading him, one cannot help but wonder about Bombelli's undaunted concentration on the sheer mechanics of these roots. In this, Bombelli is a thoroughgoing modern.[18] We of the twenty-first century are at home with algorithms, axiom systems, logical machines, software, formal structures. Just give us unequivocal rules, which prescribe a mode of operation—we ask for no more than that. Just give us the language, the symbols, the rules for combination, and we're off like demiurges, generating all allowed combinations of these symbols, producing all the structures deducible from our rules. Algorithms are so much part of our common language that the biker's T-shirt with the laconic message

DRIVE

EAT

SLEEP

REPEAT

needs no exegesis.

More to the point, all this algorithmic activity sits well in our intuitions: we usually do *not* feel that we lack the imagination to fully encompass this activity—our imaginary forces are quite up to this task. But in the later literature, Bombelli notwithstanding, we sense a continued mistrust of the use of imaginary numbers and of "new types of roots." The task of my book is to re-create the tension of imagination that yearns for a dissolution of that mistrust, and to experience the emergence of the viewpoint that enabled people to incorporate quantities like $\sqrt{-1}$ in their work and to do this with mental ease rather than mental torture.

34. Numbers as algorithms.

In discussing the "easy" case in which the indicator $d = c^2/4 - b^3/27$ is positive, I said that the manner in which Dal Ferro's expression is written tells us how to compute it (extract, as indicated, the roots and make the arithmetic operations requested by the formula). The expression doesn't provide a specific method for the extraction of those roots, but once we have such a method, the expression is itself interpretable as a possible *algorithm* for the production of a real number. It is often the case that our expressions for specific numbers suggest algorithms, or partial algorithms, for their computation. To take a random example, the number $2^{21} - 1$ happens to equal $7 \times (300{,}000 - 407)$, and this number written in decimal notation is 2097151. Each way of writing this number hints at a specific strategy for its calculation (e.g., if you express the number as $2^{21} - 1$, the form of this expression bids you do what it tells you to do to calculate the number: raise 2 to the twenty-first power and then subtract 1 from the result).

But when the indicator is negative, Dal Ferro's formula does not seem to lead us, in general, to any algorithm for computing (even approximately) the solutions of the equations it purports to solve. Bombelli's puzzle, then, is to interpret Dal Ferro's formula appropriately when the indicator is negative so that it does allow us to compute, in general, the solutions to our cubic equations.

Hints of where to look for the answer can already be found in Bombelli's treatise. As we mentioned in section 33, clearer hints can be found in the work of Viète a generation later. An even more complete solution of Bombelli's puzzle is to be found in a treatise by Albert Girard, published in 1629, titled (in translation) *New Invention in Algebra: As Much for the Solution of Equations as for Recognizing the Number of Solutions That They Have with Several Other Things Necessary for the*

Perfection of This Divine Science.[19] But a satisfactory (general) understanding of the geometry underlying Bombelli's puzzle is given only a century and a half later with the work of Abraham De Moivre and the emerging geometrization of imaginary numbers, as I will discuss in chapter 11.

35. The name of the unknown.

I mentioned, in connection with Viète's use of the term *species* for the "unknown," that we moderns refer to the unknown quantity in our algebraic equations by a letter. Traditionally, X is our letter of choice, but any convenient letter will do. Powers of the unknown, X, are indicated by decorating X with the appropriate exponent. So the fifth power of the unknown is denoted X^5.

The earlier standard word for "unknown" in texts written in Latin is *res* ("thing") and in Italian *cosa* (also meaning "thing").[20] Bombelli does not use *cosa* to refer to the unknowns he seeks in his equations, but calls his unknowns *tanti*, "amounts" or "quantities." This is his way of insisting on the concrete numericity of the solutions to his equations; he wants them to be *quantities* even when they involve the square roots of negative numbers. Here is Bombelli, spelling this out:

> If you think about it you will see that it is much more convenient to use the designation "Amount" [Tanto] than "Thing" [Cosa] because "Amount" appropriately designates a numerical quantity which is

> something you cannot say about "Thing" which is a designation applicable universally to every substance.[21]

Bombelli's notation for his unknowns is quite suggestive. He denotes them by shallow bowls. If he wishes to refer to the square, say, of the unknown, he puts a little 2 floating in the bowl, and similarly with other powers. Our X^3, for example, would appear in Bombelli's equations this way:

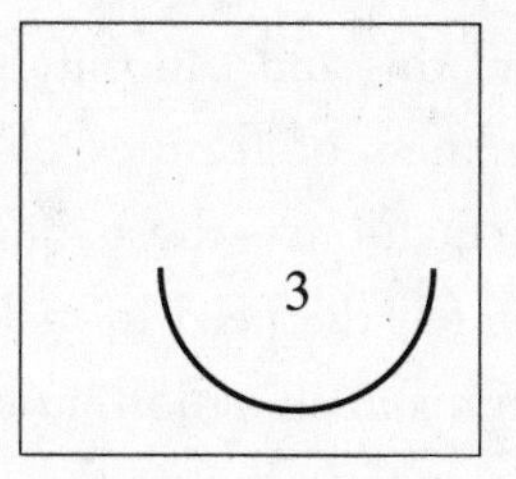

The bowl is waiting, it would seem, to be filled by the *quantity*, which will then be cubed. So it is not unusual to see in his text expressions like

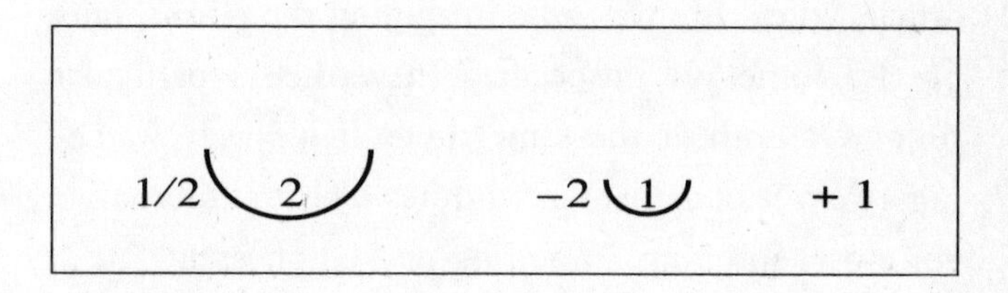

meaning, in modern terms, $X^2/2 - 2X + 1$.

The twelfth-century mathematician Bháskara reserved a somewhat stranger choice of words for the unknowns in his equations, if more than one unknown occurred in a given context. He referred to the "first" unknown with the initial syllable (*yá*) of the Sanskrit word *yávat-távat* (which means "quantity"), but the remaining unknowns he called *colors*. The symbols he used to designate them were the first syllables of the various Sanskrit words for the different colors.[22] (It is as if we restricted the letters we used to represent unknowns in our equations to *ROYGBV*.) I wish I knew why Bháskara did this and can only speculate that to counter the blankness of features of the various unknowns in his equations—those quantities without, yet, qualities—he felt the need to make them vivid by dressing them, in his imagination, in colors—the color yellow, for example.

36. Species and numbers.

In the phrase "the yellow of the tulip," there is the usual delicate ambiguity hidden in the repeated definite article, "the." Are we to be imagining the *general tulip* or else some yet unspecified, nevertheless particular tulip? Or both at the same time? If we were watching a *Nature* documentary and heard that authoritative voice-over intoning, "The ring-necked pheasant . . . ," there would be no ambiguity about that definite article:

we are in the presence of the *generic* ring-necked pheasant, whatever that is.

In algebra, however, a fruitful ambiguity surrounds the way in which one thinks of the unknown *X*. Are we thinking of *X* as a placeholder (and nothing more) for yet unspecified, nevertheless particular values? Are we thinking of *X* as a *universal value*, whatever that means? Are we thinking of *X* as a freestanding object to be treated in its own right, yet capable of being substituted (legal tender) for specific values?

Bombelli's "cup" notation that we just discussed would cleanly cast the unknown in the role of placeholder. Nevertheless, all three views of *X* play through the texts of the time. In the early eighteenth century, *Algebra* (the science of *species*) sometimes was referred to as *Universal Arithmetic*, as in the opening paragraph of Isaac Newton's little treatise on algebra:

> COMPUTATION is either perform'd by *Numbers*, as in Vulgar Arithmetick, or by *Species*, as usual among Algebraists. They are both built on the same Foundations, and aim at the same End, *viz. Arithmetick* Definitely and Particularly, *Algebra* Indefinitely and Universally; . . . But Algebra is particularly excellent in this, that whereas Arithmetick Questions are only resolv'd by proceeding from given Quantities to Quantities sought, Algebra proceeds in a retrograde Order, from the Quantities sought, as if they

> were given, to the Quantities given, as if they were sought, to the end that we may some way or other come to a Conclusion or Equation, from which one may bring out the Quantity sought. And after this Way the most difficult Problems are resolv'd the Resolutions whereof would be sought in vain from only common Arithmetick. Yet Arithmetick in all its operations is so subservient to Algebra, as they seem both but to make one perfect *Science of Computing.*[23]

The precision of this intellectual project—algebra as "universal arithmetic"—so elegantly framed by Newton in this extract, is in stark contrast to the grandness, and vagueness, of the earlier projects of universal languages, of universal symbolisms, of (following Leibniz) *Characteristica Universalis,* a calculus covering all human thought, in which controversy would be replaced by calculation. Leibniz sought "a kind of alphabet of human thoughts, i.e., a catalogue of summa genera . . . such as *a, b, c, d, e, f,* out of whose combination inferior concepts would be formed,"[24] an ambition that finds, perhaps, its wonderful, but somewhat more modest, fulfillment in formal logic and the formal languages of computers.

One of the great differences between the two ambitions, *Universal Arithmetic* and *Characteristica Universalis,* lies in the different uses of the adjective *universal.* In the first case, the *universe,* meaning the allowed range of values for which the unknown can be substituted, is *known*. In the interesting instances where we are called upon to extend the universe, to broaden the known range—as in the problems on cubic polynomials dealt with by Bombelli—we do so only with the utmost scrutiny. In the second case, the *universe* is the universe of all possible thought, an uncharted range if ever there was one.

8

STRETCHING THE IMAGE

37. The elasticity of the number line.

Let us return to the number line

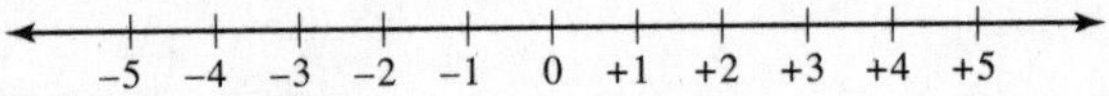

−5 −4 −3 −2 −1 0 +1 +2 +3 +4 +5

and do some experiments on it. Pick a number N, which we will be using in an exercise (in visualizing an operation with numbers). It can be any real number, but I will start off the discussion of these experiments by choosing simple whole numbers.

Let us imagine what happens when you "transform" the number line by systematically multiplying every number on the line by your chosen number N. To get a sense of this, suppose, for example, that the number you chose was +2. So the transformation consists of *doubling* each number.

Multiplying each number by 2,

$$\begin{array}{ccc} 0 & \rightarrow & 0 \\ +1 & \rightarrow & +2 \\ -1 & \rightarrow & -2 \\ +2 & \rightarrow & +4 \\ -2 & \rightarrow & -4, \end{array}$$

can be visualized as a transformation that stretches the number line like a rubber band, so that the interval from 0 to 1 now covers the interval from 0 to 2, the interval from 0 to −1 now covers the interval from 0 to −2, and so forth:

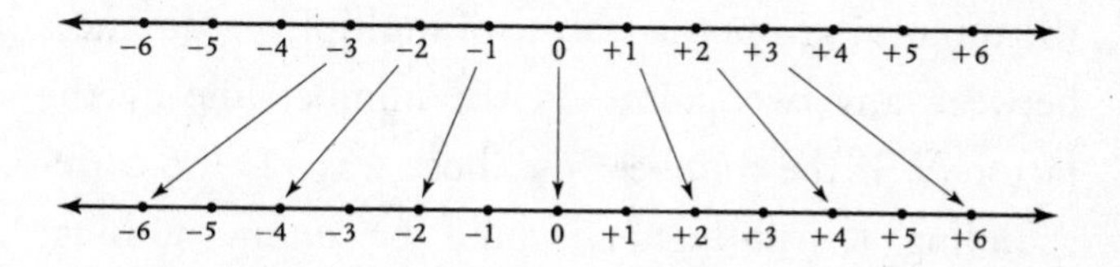

Of course, if the number you chose was +3, then the corresponding transformation would be tripling each number, and can be visualized as stretching the (quite elastic) number line so that any interval now covers the length of three intervals:

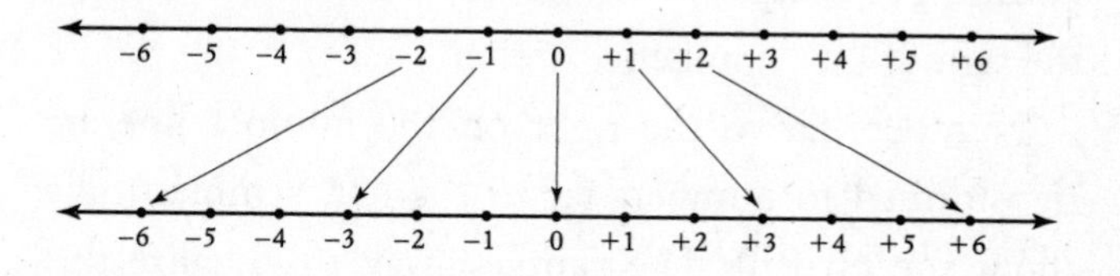

This act of *multiplying by a fixed number* is something that can be done, and visualized, for any number N, not only whole numbers. If N is a positive number, then the transformation can be thought of as simply a *zoom*, a *change of scale.* If N is bigger than 1, then the transformation stretches the number line; if N is less than 1, then it compresses the line. For example, if your number line is measured off in marks that occur every inch and the transformation you subject it to is multiplication by $N = 0.3937\ldots$, then the effect is to produce a line measured off in marks that occur every centimeter.

In general, multiplying by a positive number N has the uniform geometric feature of multiplying distances between any two points on the number line by the factor N. If the number you chose was +1, the corresponding transformation would be doing nothing: leaving everything utterly unmoved. We are now ready to try to visualize what happens if the number you chose was −1.

Multiplying a positive number by −1 simply changes the sign of the number. But, as we have discussed in part I, the product of two negatives is positive: multiplication of a negative number by −1 also just changes the sign of that number.

Numbers far to the right on the number line are transformed to numbers far to the left, symmetrically about the pivot 0. The simplest way to visualize this transformation, *multiplication of each number by* -1, is that it flips the number line about 0; or, if you wish, rotates the line (on the page) around 0 by 180 degrees. The number line does an "about-face." We might imagine this operation performed by sticking a pin through the point 0 and rotating the number line 180 degrees, in the plane of the page, around the fixed point 0:

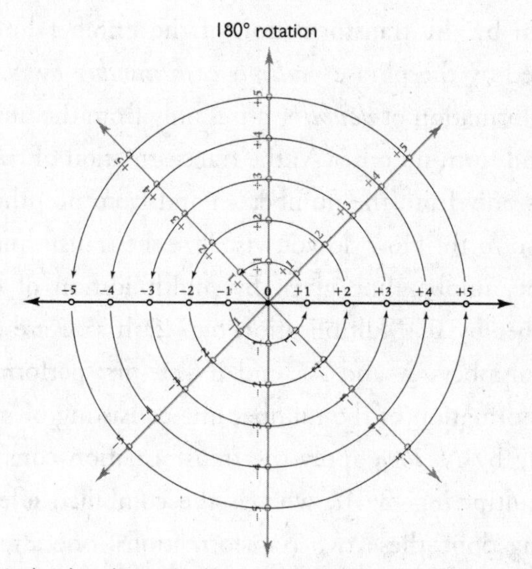

Multiplying by -1 can be visualized as rotating the number line 180 degrees, keeping 0 fixed.

Similarly, each real number N, positive or negative (or even zero), gives rise to its own specific transformation of the number line (multiply each of the numbers on the line by N):

$$\begin{array}{ccc} 0 & \rightarrow & 0 \\ +1 & \rightarrow & +N \\ -1 & \rightarrow & -N \\ +2 & \rightarrow & +2N \\ -2 & \rightarrow & -2N \\ +3 & \rightarrow & +3N \\ -3 & \rightarrow & -3N. \end{array}$$

Conversely, each such number N is uniquely determined by the transformation of the number line described by the phrase *multiply each number by N*. The transformation of *doubling* arises only from the number +2 and from no other N; the transformation of *tripling* arises only from the number +3 and from no other N; and so forth. How do you visualize the transformation of the number line given by multiplication of every number by 0? Multiplication by −2? If you are given *two* numbers N and M, and if you *first* perform the transformation of the number line consisting of multiplying by N, *then* apply the transformation consisting of multiplying by M, what is the combined effect of having done these two transformations, one after the other, on the number line? (Answer: You have multiplied by the product of the two numbers, $M \times N$.) The transformation of the number line described by the phrase *multiply each number by N* is a way of representing geometrically the role that the number N plays in the process of multiplication.

"Representing geometrically" is an innocent phrase, but it suggests a significant shift in our attitude toward *number*, a shift that is of great importance if we are satisfactorily to imagine imaginary numbers. If we are to do so, we must first be capable of thinking of *numbers* geometrically.

To underscore this shift in attitude, I would like to push it to the extreme, but if you object too strenuously, leave it undone for now. Let us play with the idea of simply *identifying* the number N with the geometric transformation of the number line described by the phrase *multiply each number by N.*

Can we regard, for our present discussion, a given number N as just "being" that unique transformation of the number line described by the phrase *multiply each number by N*? For instance, can we think of the number 2 as the *act of doubling* (numbers on the number line)? Or the number 3 as the *act of tripling*?

Return to the number -1, whose geometric representation is rotation of the number line by 180 degrees. When you perform this rotation you've done a pirouette and you're back where you started. Which figures, because

$$-1 \times -1 = +1.$$

If you find this a novel way of imagining numbers, as "acts," you are not alone. It may seem peculiar, especially to people who prefer to regard *number* as a settled notion, that we are asking permission to think of numbers, now, as *acts*. In the history of the concept of number, *number* has been *adjective* (*three* cows, *three* monads) and *noun* (*three*, pure and simple), and now, with the "identification" we wish to make, *number* seems to be more like a *verb* (*to triple*). But we shouldn't be surprised by these protean shifts, for surely the biography of number, even in the twenty-first century, is only in its infancy.

38. "To imagine" versus "to picture."

Despite the etymology of the word *imagine* that we recalled earlier (see sect. 2), to *imagine* an (imaginary) object and to *visualize* it can be very different activities. Sometimes, of course, they are the same. For example, when I suggested (in sect. 1) that we *imagine* an elephant, and helpfully provided a box for it, clearly our imaginative faculty was called upon to conjure up a picture, and no more: the iconic shape of an elephant with its inventory of trunk, floppy ears, four stumplike legs, and delicate tail. The exercise was to imagine the shape of an elephant. It would be similarly easy, I suppose, if we were asked to imagine the feel or the smell of an elephant. The exercise was not, however, to imagine *being an elephant*, which might have been harder.

To *visualize* something that we have the mental equipment to visualize, we *will* a picture onto the mind's screen. This may be a tricky activity, but at least it is the type of activity we have done before. If we set about to *imagine* something that resists the types of mental image-making activities we are familiar with, we must work at new ways of providing a natural home for that something in the mind's mind.

The objects and forms invoked in poetry and literature must be rigorously imagined, but whether they must, or can, be directly visualized is a trickier question. When *The Metamorphosis* was to be published as a book, Franz Kafka's publisher wrote to him suggesting that the cover illustrator "might want to draw the insect itself." Kafka wrote back:

> Not that, please not that! . . . The insect itself cannot be depicted. It cannot even be shown from a distance.[1]

Vladimir Nabokov, in his writings and lectures on *The Metamorphosis*, had a different idea about whether or not the insect should, or could, be depicted. Consider, for example, Nabokov's own architectural sketch of it:

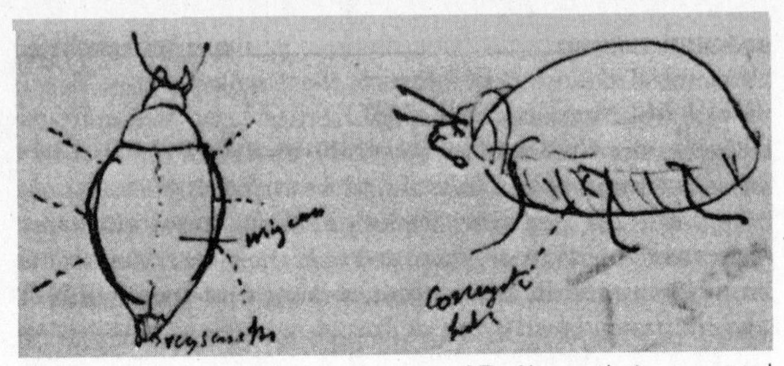

From the first page of Nabokov's teaching copy of *The Metamorphosis*, as annotated by himself. (From V. Nabokov, *Lectures on Literature*, ed. Fredson Bowers [Harcourt Brace, 1980], p. 250.)

But, I believe, Kafka's point is excellently well made precisely by Nabokov's further reflections on the geometry of the insect. Nabokov, having convinced himself that the creature is a beetle, having noted that his "hard round back is suggestive of wing cases," and having mused that "Gregor the beetle never found out that he had wings under the hard covering of his back," has then to reconcile the precision of his own picture ("the precision of poetry and the intuition of science," as Nabokov had written elsewhere) with the further information, provided by Kafka, that Gregor had

> tried at least a hundred times, shutting his eyes, to keep from seeing his wriggly legs, and only desisted when he began to feel in his side a faint dull ache he had never experienced before.[2]

Nabokov then points out a structural problem: "a beetle has no eyelids and therefore cannot close its eyes—a beetle with human eyes." And to resolve it, he writes:

> [Gregor] is half-awake—he realizes his plight without surprise, with a childish acceptance of it, and at the same time he still clings to human memories, human experiences. The metamorphosis is not quite complete as yet.[3]

This brings home the sense that, despite (or better still, because of) Kafka's clarity, a kind of imagination is called for here that goes beyond architectural visualization.

A movie of Gogol's story "The Nose," in which a disembodied nose goes "driving about all over town under the guise of a State Councillor," could quite easily miss the point of the tale. For Gogol in his writing is prodding us, cajoling us, instructing us to engage in precisely the exercise of imagining this unvisualizable image in the full glory of its unvisualizability.[4]

In her essay "Imagining Flowers," Elaine Scarry emphasizes how hard it is to imagine a face, in contrast to a flower. She writes: "[T]he daydreamed face expresses the lapse of the imagination from the perceptual ideal."[5] But the problem of recalling the faces of absent friends may be that our imagination of their presence, their own inner life, and their dynamic relationship to us is so vivid that we find it difficult to *suppress* these imaginings, to focus on the mere static visual image-memory of a face. But there is no such problem with flowers. How much inner life, after all, does a tulip have?

The act of visualization is, to be sure, only one possible act in the repertoire of the imagination. To visualize, we play the image on our already existing internal screen. But the more difficult leaps of the imagination force us to establish larger screens and, perhaps, new theaters of the mind.

For example, the exercise we are engaged in, to imagine imaginary numbers, is not a simple act of visualization. Rather, we will try to do this in two steps:

- first, to comprehend the idea of *number* as *transformation*, and
- second, to work at visualizing these transformations.

39. The inventors of writing.

John Ashbery, in his prose poem "Whatever It Is, Wherever You Are," has put before us an intricate exercise. We are asked to imagine our ancestors, the inventors of writing, who are imagining us (the present readers, and the poet):

> Probably they meant for us to enjoy the things they enjoyed, like late summer evenings.

This conceit, of course, is decidedly not set in any historical frame. The "they," for example, are not pinned down as scribes in some Sumerian city-state of the third millennium B.C.E., entranced by the imagined power of their invention of cuneiform. Ashbery's word *probably* sets us squarely in our poet's workroom, deliberating the probable intentions of his creatures "the inventors." We see the poet and his somewhat tentative stance toward his own imaginative construction. But this will change fast in Ashbery's poem: the images will pile on faster, etched with a surer hand, even as writing will show itself to be less and less graspable.

Some poetic constructions stride onto the page fully formed from the beginning. But some begin tentatively, as Ashbery's does, as if they are not yet totally seen, but gather solidity, determinateness, only as they are recounted.*

Skipping a line in Ashbery's prose poem, one reads:

> Singing the way they did, in the old time, we can sometimes see through the tissues and tracings the genetic process has laid down between us and them. The tendrils can suggest a hand; or a specific color—

*Consider the appearance and then disappearance of the word *likeness* in the description of the details of Ezekiel's vision (as in Ezek. 1.22, 23):

> 22. And the likeness of the firmament upon the heads of the living creature was as the colour of the terrible crystal, fetched forth over their heads above.
> 23. And under the firmament were their wings straight . . .

> the yellow of the tulip, for instance—will flash for a moment in such a way that after it has been withdrawn we can be sure there was no imagining, no auto-suggestion here, but at the same time it becomes as useless as all subtracted memories.

The first sentence of this extract is indeed grammatical, although its grammar shimmers in a maddening way until the verb *see* is seen to be used intransitively, and the effort of reading the sentence is seen to reenact its message. The word *tendrils* refers to the cursive (I imagine) script on the page that, as it winds around the line, can suggest one thing (a hand) as easily as another (a color).

"The tendrils can suggest a hand" would seem to portray this new invention as a gentle, helpful agent, kindly offering images to us, but by the end of that sentence, we get a whiff of its hallucinogenic potency. Ashbery uses the verbs *withdraw* and *subtract* in something of a parallel construction, and we are left to wonder how to work out that parallel. But the overall sense is that we have little control over this new invention, which leaves us, in the end, only with things "as useless as all subtracted memories."

Now let us return to the mathematical setting and try to control $\sqrt{-1}$, even though we do not yet have a graspable picture of it.

40. Arithmetic in the realm of imaginary numbers.

Ignore the fact that we have, as yet, no image of $\sqrt{-1}$, and take for granted the one defining characteristic of $\sqrt{-1}$, namely, that its square is -1. This strange quantity $\sqrt{-1}$ is traditionally referred to by the symbol i, and it may clarify our thinking if we refer to $\sqrt{-1}$ by this "proper name," i, as we review the arithmetic laws that it satisfies.*

First, and most important, i has the basic property that

$$i^2 = i \times i = -1.$$

If $-A$ is a negative real number, then we will interpret an expression of the form $\sqrt{-A}$ (a quantity Bombelli would refer to as *più di meno*!) as

$$\sqrt{-A} \quad = \quad \sqrt{-1} \cdot \sqrt{A} \quad = \quad i \cdot \sqrt{A} \quad = \quad \sqrt{A} \cdot i,$$

where $\sqrt{A}$ means the positive square root of the number A. For example,

$$\sqrt{-2} = 1.414 \ldots \cdot i.$$

Second, if we are serious in thinking of i as a quantity, we should allow ourselves the liberty of considering more general expressions like $5.3 + (6.1)i$, which can also be written as

*If you are puzzled about why a simple change of notation of this sort can clarify any thinking, see section 48.

$$5.3 + 6.1\sqrt{-1}$$

or

$$5.3 + \sqrt{-1 \times (6.1)^2}$$

or

$$5.3 + \sqrt{-37.21}$$

as quantities. To be systematic about it, then, let us formulate laws to deal with all expressions of the form $a + bi$ where a and b are real numbers.

At this point we should pause to do a bit of terminological housekeeping. We use the term *imaginary number* to refer only to a multiple of i by a real number (or equivalently, to refer to the square root of a negative real number), and we use the term *complex number* to refer to the more general expression $a + bi$, where a and b are real numbers. Any complex number as displayed above is uniquely expressible as a sum of a real number a (called its *real part*) and an imaginary number $bi = b\sqrt{-1}$ (called its *imaginary part*). (According to this, the number zero has the honorary position of being both real and imaginary, which sort of figures!) We can write $a + b\sqrt{-1}$ also as $a + bi$ or $a + ib$, as the mood takes us.

Addition of complex numbers follows a simple rule: to add two complex numbers, $P = a + bi$ and $Q = c + di$, just add their real parts separately and their imaginary parts separately, and then put these together, giving

$$P + Q = (a + c) + (b + d)i = (a + c) + (b + d)\sqrt{-1}$$

as in

$$\begin{array}{rcl} 5.3 & + & 6.1\sqrt{-1} \\ +2.2 & - & 3.1\sqrt{-1} \\ \hline 7.5 & + & 3.0\sqrt{-1} \end{array}$$

The law for *multiplication* of complex numbers obeys the natural extension of the *distributive law* (discussed at such length in part I): to multiply two "quantities" $a + bi$ and $c + di$, where a, b, c, d are real numbers, you just follow this rule (noting here that $bi \times di = (bd)i^2 = -bd$):

$$(a + bi) \times (c + di) = ac + adi + bci - bd.$$

Or, collecting terms, the product of the two complex numbers $a + bi$ and $c + di$ is equal to the complex number:

$$(a + bi) \times (c + di) = (ac - bd) + (ad + bc)i.$$

For example, in multiplying

$$(5 + \sqrt{-3}) \times (3\sqrt{2} + \sqrt{-1}),$$

which can also be written as

$$(5 + \sqrt{3}i) \times (3\sqrt{2} + i),$$

you get four terms, corresponding to multiplying the 5 by the $3\sqrt{2}$ and the i, and multiplying the $\sqrt{-3} = \sqrt{3}i$ by the $3\sqrt{2}$ and the i. Finally, you add the four terms together, and this gives

$$(5 + \sqrt{3}i) \times (3\sqrt{2} + i) = (15\sqrt{2} - \sqrt{3}) + (5 + 3\sqrt{6})i.$$

In summary, we can add complex numbers and we can multiply them.

If you agree to this, we are ready to perform another experiment. It is an especially valuable experiment if you have had *absolutely no experience* with mathematics.

Cube the quantity

$$\frac{1 + \sqrt{-3}}{2}.$$

If we use the symbol i in expressing this quantity, we could also write the above expression as

$$\frac{1 + \sqrt{-3}}{2} = \frac{1 + \sqrt{3}i}{2} = \frac{1}{2} + \frac{\sqrt{3}i}{2}.$$

So, the problem is to multiply out:

$$\frac{1 + \sqrt{-3}}{2} \times \frac{1 + \sqrt{-3}}{2} \times \frac{1 + \sqrt{-3}}{2}$$

or, writing this with the symbol i, multiply

$$\left(\frac{1}{2} + \frac{\sqrt{3}i}{2}\right) \times \left(\frac{1}{2} + \frac{\sqrt{3}i}{2}\right) \times \left(\frac{1}{2} + \frac{\sqrt{3}i}{2}\right).$$

Do the computation carefully, on paper, using the rules we agreed to.* *Then* ponder your answer. Just do it!

When you get your answer, which should be something of a surprise to you, don't stop there. Think of what your answer might possibly mean or might imply.

41. The absence of time in mathematics.

In learning mathematics, there is sometimes a clear sequence of steps, one following upon another, that you must go through to understand the math. Missing a step, or doing things out of step, is not advisable. This is one of the underpinnings of Descartes's *Rules for the Direction of the Natural Intelligence.*[6] You must *first* understand step 1 and *then* step 2.

When you read the word *then* in a historical text (e.g., "He then proceeded to state that the guarantee he had given Czechoslovakia no longer in his opinion had validity"),[7] that word can have the effect of framing the present action within a system of time-ordered events. But the word *then* in mathematics ("If X then Y") is not an elementary chronometer. It tends to mean "therefore," as it partly does in the first line of Shakespeare's sonnet 90, which plays on the words *then*, *when*, and *now*:

*The straightforward way to do this is to first multiply $(1 + \sqrt{-3})/2$ by $(1 + \sqrt{-3})/2$ and then multiply the product by $(1 + \sqrt{-3})/2$.

Then hate me when thou wilt, if ever, now,
Now while the world is bent my deeds to cross.

To give the mathematical *then* any temporal significance at all, one might interpret it, for example, as fixing our present position on a somewhat subjunctive epistemological time line: "If (at a given time) I knew *X*, then (shortly afterward, by such and such an argument) I would know *Y*." Apart from this epistemological development, there is no temporal "before" and "after" in the *logical* structure of mathematics. The lines in Shakespeare's sonnet 38

And he that calls on thee, let him bring forth
Eternal numbers to out-live long date.

(where *numbers* could mean "verses" but also "numbers") play on the weirdness of considering notions of temporality in the sphere of number.

In contrast, it would be hard to write about writing, as Ashbery has done ("Singing the way they did, . . ."), without evoking at least two moments of time, explicitly tagged: the time when the act of writing occurred and the time when the act of reading it occurred. Ashbery has added an extra zigzag of time references in his account, for we *now* are being asked, in his prose poem, to think of the inventors of writing *then* who were thinking of us *now* reading what they are *now* writing—or, should I say, were *then* writing? For the sense of any firm present time has been artfully obliterated in Ashbery's account. But even this shuttling over the time line is less complex than in Shakespeare's sonnet 17, which begins

> *Who will believe my verse in time to come*
> *If it were filled with your most high deserts?*

For one quickly learns that the verse referred to is, in fact, not yet written:

> *If I could write the beauty of your eyes,*
> *And in fresh numbers number all your graces,*
> *The age to come would say, "This poet lies;*
> *Such heavenly touches ne'er touched earthly faces."*
> *So should my papers (yellowed with their age)*
> *Be scorned like old men of less truth than tongue,*
> *And your true rights be termed a poet's rage*
> *And stretchèd meter of an antique song.*

42. Questioning answers.

Have you cubed $(1 + \sqrt{-3})/2$? The correct answer is -1. That is,

$$\left(\frac{1 + \sqrt{-3}}{2}\right)^3 = -1.$$

There is, in fact another number, easier to guess, whose cube is -1, and yet a third number. Can you find them?

To check that you have the right answers, consult this endnote.[8]

There are also three numbers whose cube is +1. One of those three numbers is +1 itself. The other two are

$$\frac{-1-\sqrt{-3}}{2} \text{ and } \frac{-1+\sqrt{-3}}{2}.$$

Check this by cubing each of these numbers.

We can phrase the results of these experiments this way: the number −1 has *three* cube roots, as does the number +1, and we know what they are.

Whenever we get an answer to any mathematical question, it is a good idea to turn the answer around and question it. What does it signify? How can we use the answer to go further?

There are judicious moments to pause in our work, to take a break. But now is not such a moment. Let us resist the temptation to stop thinking about our little calculation just when we have succeeded in executing it. Let us, rather, ask the answer to this calculation to help us in the next tier of questions. How can the three cube roots of −1 enlighten us?

43. Back to Bombelli's puzzle.

We have learned from our experimentation that the number −1 has

$$(1 + \sqrt{-3})/2 = 1/2 + \sqrt{-3/2})i$$

as a cube root, and it has two other cube roots as well! So if we are ever to dare use the symbol $\sqrt[3]{-1}$, we must be careful to realize that it is (triply) ambiguous. We might be meaning

$$\sqrt[3]{-1} = \frac{1}{2} + \left(\frac{\sqrt{3}}{2}\right)i$$

or

$$\sqrt[3]{-1} = \frac{1}{2} - \left(\frac{\sqrt{3}}{2}\right)i$$

or, less impressively,
or, less impressively,

$$\sqrt[3]{-1} = -1.$$

With this in mind, let us return to our discussion in section 33, where we looked at the cubic equation

$$X^3 = 3X - 2$$

and noticed that both $X = 1$ and $X = -2$ are solutions (and are the only ones) and that (since the indicator of the equation is 0, and $c = -2$) the oracular Dal Ferro expression gives as its solution

$$X = \sqrt[3]{-1} + \sqrt[3]{-1}.$$

Well, given the triple ambiguity of each $\sqrt{-1}$ in this expression, there seems to be a sixfold[9] ambiguity in their sum! Can you find *both* actual solutions ($X = 1$ and

$X = -2$) among them? Experiment before looking at the endnote for the answer.[10]

Bombelli did come to believe that his cubic radicals, or at least the expressions given by Dal Ferro's formula, "exist"; but why did he believe this, and what did their existence mean to him?

44. Interviewing Bombelli.

I am not trained as a historian and so cannot easily see through the tissues and tracings of the early Italian algebraists. In thinking about the first treatises on these imaginary numbers, I find myself daydreaming about what mental picture they (Cardano, Bombelli, or their colleagues) could conceivably carry along.

Here is a way of making this daydream more specific. Imagine that you are allowed to interview a mathematician of the old time, let us say Bombelli, who has "accepted" (in whatever sense can be meant by this term) numbers like $(1 + \sqrt{-3})/2$ and who is working through the question of "existence" of his cubic radical expressions. How would he respond to you, an "investigative journalist," if you wanted to probe further?

You: Signor Bombelli, since you are comfortable dealing with monstrosities like $(1 + \sqrt{-3})/2$ and your cubic radicals, do you also accept further elaborations of these numbers, like

$$\sqrt[5]{\sqrt{2} + \sqrt{-1}}?$$

Bombelli: Ma sì! L'arte di trattare le quantità immaginarie è sottile ma duttile. Perché mai però si prende la briga di chiamare in causa una simile complessità? [But yes! The art of dealing with imaginary quantities is subtle but supple. But why on earth would you bother to call forth such a complexity?]

You: But is this

$$\sqrt[5]{\sqrt{2} + \sqrt{-1}}$$

yet a "new" kind of radical? Or is it, perhaps in disguise, something you have already dealt with in your treatise?

Bombelli: Non importa! Io *so* come trattarlo, esattamente come ho fatto, con discreto successo, con $\sqrt{-1}$. Ad esempio, posso sommare o moltiplicare il suo disgraziato

$$\sqrt[5]{\sqrt{2} + \sqrt{-1}}$$

per altre brutture similmente odiose (sono restio a chiamarle "quantità"), semplicemente usando le regole comuni dell'algebra— [No matter! I *can* deal with it, exactly as with $\sqrt{-1}$, or as with my cubic radicals. For example, I can add and multiply your wretched

$$\sqrt[5]{\sqrt{2} + \sqrt{-1}}$$

with other similarly odious concoctions (I hesitate to call them "quantities") just by using the standard rules of algebra—]

You: And are there worlds without end here? Would you accept even further "new" quantities by taking roots of sums of roots of things like

$$\sqrt[7]{\sqrt[5]{2 + \sqrt{-15}} + \sqrt[3]{2 - \sqrt{-15}}} + \sqrt[7]{\sqrt[5]{2 + \sqrt{-15}} + \sqrt[5]{2 - \sqrt{-15}}}$$

and, each time, finding more and more—

Bombelli: Sparisca! [Get out!][11]

The question of whether there are worlds without end, whether extraction of roots of sums of roots leads to more and more complicated proliferation of "sophistic quantities," might well have perplexed Bombelli.[12]

The old-time mathematicians weren't the only ones who had trouble understanding collections of things that behave like numbers. There are more than a few living mathematicians, I would guess, who can empathize with Bombelli's plight of working, perhaps fairly successfully, with systems of numbers whose logic is unassailable, but for which an immediately graspable explanatory picture is not yet available.

9

PUTTING GEOMETRY INTO NUMBERS

45. Many hands.

In this book we have been shuttling between two vastly different mental experiences:

- the generation of a single visual image in our minds as we read lines of verse ("a hand, or a specific color—the yellow of the tulip, for instance—"), written, one may assume, by a single author; and
- the cultivation of a comprehensive inner intuition for *imaginary numbers*, the fruit of collective imaginative labors over time.

The mathematical idea did not spring from one imagination. Nor was it, in any simple sense, a communal effort, either. Nor is there any definitive text upon which we are obliged to rely in order to achieve our understanding of the idea.

Searching for a parallel in literature for the manner in which this mathematical intuition developed, might we look to the understanding of the nature of epic poetry propounded by Milman Parry and A. B. Lord?[1] The Homeric tradition had many singers, many variant songs, with further variations from performance to performance, long before the *Iliad* or *Odyssey* was written down. And (following G. Nagy) even when those songs were eventually written, they had many texts. Or, to pass to a different tradition, consider the tale told in the Iranian *Book of Kings* (Ferdowsī's *Shāh-nāmeh*) describing its own creation: a myth of a complex synthesis of multiple oral and written authorships. Here it is, as recounted in Nagy's book *Homeric Questions*: "A noble vizier assembles *mōbads*, wise men who are experts in the Law of Zoroaster, from all over the Empire, and each of these *mōbads* brings with him a 'fragment' of a long-lost *Book of Kings* that had been scattered to the winds; each of these experts is called upon to recite, in turn, his respective 'fragment,' and the vizier composes a book out of these recitations."[2]

But the short answer to the rhetorical question asked above—Can we look to epic poetry's tradition for a parallel?—is: Not exactly. Each intellectual community has its own way of developing its ideas, in and outside its written records, and its own way of handing them down. The specific ways that mathematical truths move from person to person, and how they are transformed in the process, are as difficult to capture as the truths themselves.

46. Imagining the dynamics of multiplication by $\sqrt{-1}$: algebra and geometry mixed.

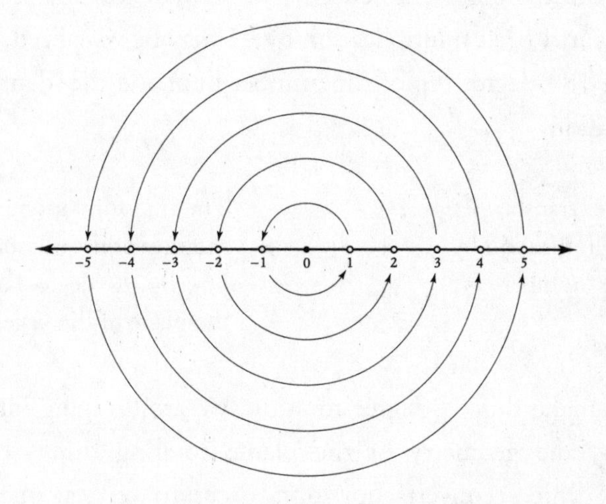

If multiplication by −1 is the rotation of the number line by 180 degrees, and if you have agreed to identify the number −1 with this act, you are ready for the revelation that emerged in written form only at the turn of the nineteenth century. I hope, in fact, that you have already seen this on your own.

One way to *realize* the number $i = \sqrt{-1}$, to make it real for ourselves, is to identify this i with the transformation *rotation of the number line by 90 degrees in the plane of the page.* For if we can truly lodge this in our imagination, then the defining property of *i*, that is, the fact that $i \times i = -1$, becomes eminently visualizable: rotate by 90 degrees (counterclockwise, let us say) in the plane of the page, and then do it again, and you have achieved a rotation by 180 degrees.

But there is a difference between what we have seen in part I (that multiplication by −1 may be visualized as the 180-degree flip of the number line) and the identification,

The "transformation": multiplication by the "number" *i*	←------→	The transformation: rotation of the number line by 90 degrees in the plane of the page.

To make this 90-degree rotation, we are bringing into play the geometry of the plane: rotating things by 90 degrees converts horizontal lines to vertical ones, and vertical lines to horizontal ones.

If we are looking for a "place" to put this new number *i*, where can we place it so that its position is appropriate to (and explains its connection to) the transformation (rotation of the number line by 90 degrees in the plane of the page) with which we are identifying it?

Let us take time to think about this, and let's also think about the turnabout in our focus. We started with the project of thinking about *number*, the mainstay of *algebra*. At first we found ourselves led to thinking about transformations of the number line: stretching the number line or flipping it onto itself by a 180-degree rotation. But now we are dealing with 90-degree rotations, bringing into play the *geometry of the plane*.

47. Writing and singing.

For John Ashbery (as for Parry and Lord), the early writers weren't writers entirely, they were singers: "Singing the way they did, in the old time. . . ." A singer singing to a present audience is also maintaining (possibly extending) the tradition of song for other singers. For other present singers. Continuity is all. The novel element in writing is that it no longer requires such proximity. It allows for conversations, for the communication of intentions, over centuries.

> Probably they meant for us to enjoy the things they
> enjoyed, like late summer evenings, and hoped that
> we'd find others and thank them for providing us
> with the wherewithal to find and enjoy them,

writes Ashbery.[3] There is a cheerful, convivial feel to all this, making us momentarily forget the more somber tonalities elsewhere in Ashbery's text: the words *withdrawn, subtracted, useless, quicksand, bog.*

And it is at this point that we can enjoy the mild

irony in the *they* and the *us* in the sentence quoted above: the *they*, we imagine, refers to the inventors of writing; the *us* refers to their readers. But the evolution of writing, and its speciation, is continual. All writers, poets, are "inventors." All readers, too. The *they* are *us*; the *us* are *they*.

Inventors of a medium, never completely sure that their invention can actually communicate, keep testing it out. For the early explorers of amateur radio, the sum and substance of messages were often simply, "How do you read me? Over." And consider these lines of Adrienne Rich:

> *I know you are reading this poem which is not in your language*
> *guessing at some words while others keep you reading*
> *and I want to know which words they are.*[4]

48. The power of notation.

I asked earlier: Where can we place the number $i = \sqrt{-1}$ so that its position is appropriate to (and explains its connection to) the transformation *rotation of the plane by 90 degrees* with which we are identifying it? We should be aware, when we contemplate this question, that the very framing of it throws us into the late eighteenth or early nineteenth century. We are most emphatically out of the context of sixteenth-century algebra, for our question is one that was not in the vocabulary of Cardano. But however reflective we are about these matters, it seems tricky, if not impossible, to deal with early mathematics (or with any mathematics, for that matter) without helping it along by asking questions about it that extend its frame, even though they may be, in the case of early mathematics, thoroughly anachronistic.

Notation, for example, is enormously important in mathematics. A seemingly modest change of notation may suggest a radical shift in viewpoint. Any new notation may ask new questions. Most translators of early mathematics, Cardano's translator included, make use of concise modern algebraic notation to "clarify" the text, but in so doing often manage to beg some questions, answer others, and hide yet others. We have done so, too, throughout this book. Take, for example, the symbol we've been using for square root or cube root. We have said not a word about those clean angular signs,

$$\sqrt[2]{\ },\ \sqrt[3]{\ },\ \sqrt[4]{\ },\ \ldots,$$

which shelter the number whose square or cube root is to be taken, and which cradle the little 2 (though the 2 can be omitted, as we've seen) or 3 that says whether the square root or the cube root is meant. As mentioned in the introduction (sect. 1), in sixteenth-century Italian mathematics, the word *lato* (meaning "side") would sometimes signify square root, the idea being that the number whose square root is being taken is cast into the role of *area* of a square, and its square root (its *lato*) is the length of a side of that square. Similarly, cube root was referred to as *lato cubico*, which would cast the original number as the volume of a cube, with its cube root the length of any side of that cube.

Bombelli used this terminology, but it was also common, even earlier, for mathematicians to use the symbol R., or rather something closer to the universal sign for medical prescriptions, ℞ (short for *radix*), to indicate roots.[5] Bombelli denoted square root in his *L'Algebra* (most of the time) as R.q. (*radice quadrata*), so the square root of 2 would be R.q.2, and cube root he denoted R.c. (*radice cubica*).

Any seemingly harmless change of term, for example, from *lato* to the less evocative R.q., invites us to think, literally, in different terms. Using R.q. for the operation of *taking a square root* makes it, as a process, less immediately tied to its original geometric signification. This defamiliarization, putting the geometry of the process further from our imagination and rendering the process of taking a square root a *black box*,

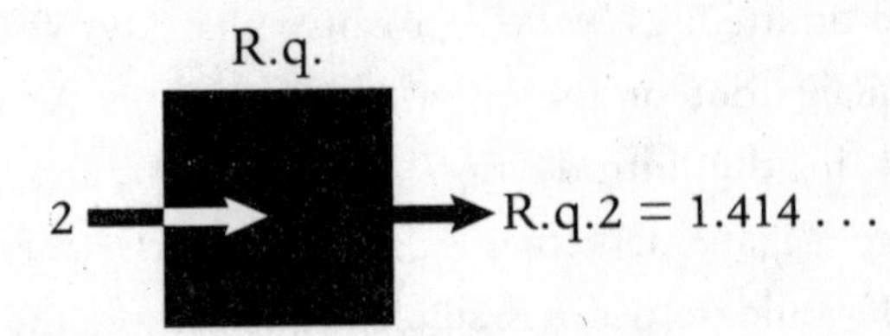

has the liberating effect of allowing us to dwell on the formal profile of the operation, and allows us to shed an imaginative but restrictive picture, opening new possibilities. The fastest way, perhaps, to make room for a new evolution of the imagination is to repress old imaginings. This is akin to the gardener plucking off the dead heads, the spent flowers, to encourage new blooms.

Perhaps our modern symbol $\sqrt{\ }$ evolved from the prescriptive ℞ , or simply from the letter *r* written cursively:

And the idea of then cradling the little 2 or 3 in the convenient crook of the "elbow" would have been natural. If either of these prescriptions is the way the square-root symbol came about, then that symbol is like a well-worn, well-used tool, the shape of its handle fitting the hand.

What can we say about a mathematical operation that has shaped its own symbol, forming a new printer's character? One thing we can say is that, as an operation, it must have achieved a certain level of familiarity. It is an old tool, a well-known one. It is something that needs less explanation, and the less it needs to be explained, the more it can serve as a way of explaining other things.

It is also easy to underestimate the difficulties of comprehension that any change of notation presents. For example, when Roman numerals were replaced with Arabic numerals, the difficulties for contemporaries faced with this shift were enormous. To get a sense of this, let us read a few lines of one of the earliest English math primers, which, in dialogue form, shows a *Master* examining a *Student* to make sure the student has grasped the new notation.

> **Master:** Write these iii nombers eche by it selfe as I speake them vii, iiii, iii.
>
> **Student:** 7, 4, 3.
>
> **Master:** How write you these four others ii, i, ix, viii?
>
> **Student:** Thus (I trowe) 2, 1, 6, 8.
>
> **Master:** Nay. There you mysse: loke on myne exa[m]ple again.
>
> **Student:** Syr trouth it is, I was to blame, I toke 6 for 9.*
>
> **Master:** Now then take hede, these certayne valewes every fygure representeth, when it is alone written without other fygures ioyned to hym. And also when it is in the fyrst place though many other do folowe, as for example, this figure 9 is ix standynge now alone.

*The typesetter of what should really be sonnet 116 in the 1609 Quarto edition of Shakespeare's sonnets might have profited from this little dialogue (the sonnet was misnumbered as 119).

> **Student:** How is he alone and standeth in the myddel of so many letters?
>
> **Master:** The letters are none of his felowes. And if you were in Frau[n]ce in the myddle of Frenceh men, yf there wer no Englyshe man with you, you would recken your selfe to be alone.[6]

The question of whether to replace *original* by *editorialized and modernized* notation is one well known to Shakespeare scholars. See, for example, the five-page note in Stephen Booth's *Shakespeare's Sonnets* on the strong and weak points (mainly weak points, according to Booth) of the argument made by Robert Graves and Laura Riding in their essay "A Study in Original Punctuation and Spelling,"[7] in which they insist upon the 1609 Quarto spelling and punctuation for Shakespeare's sonnet 129 ("TH'expence of Spirit in a wafte of fhame . . .").*

49. A plane of numbers.

I have asked my question twice before, and at the risk of sounding more and more like the opening lines of an old Jewish joke,† I would like to get back to it:

*For example, the line of the sonnet that Booth prints in its modern orthography as "A bliss in proof, and proved, a very woe" was set as "A bliffe in proof and proud and very wo."

†To tell this joke we first need a glossary. *Gimmel* is the third letter of the Hebrew alphabet, sounding like a hard *G. Rashi* is one of the prin-

> Where can we place the number $i = \sqrt{-1}$ so that its position is appropriate to (and explains its connection to) the transformation *rotation of the plane by 90 degrees* with which we are identifying it?

Whatever answer we eventually give to this question, we clearly need to be dealing with a plane within which this 90-degree rotation can be performed, and also with the number line itself. Here, then, is our inventory. We have the number line, which I will picture as an infinite ruler:

−6 −5 −4 −3 −2 −1 0 +1 +2 +3 +4 +5 +6

We want to put the number line on the plane in preparation for rotating it, and I suggest that we simply position it on the horizontal axis. In fact, let us identify the horizontal axis on the plane with the line of real numbers.

---

cipal commentators of the Pentateuch, the spelling of whose name does not contain the letter gimmel.

Here is the joke. Two Talmudic scholars are in conversation. I'll call them A and B.

A: *WHY* should Rashi be spelled with a gimmel?
B: But it isn't.
A: You are not really understanding my question: *WHY* [pause] should Rashi be spelled with a gimmel?
B: But Rashi is *not* spelled with a gimmel.
A: *Please* try to understand my question [he repeats, even more slowly]: *WHY* should Rashi be spelled with a gimmel?
B: But why *SHOULD* Rashi be spelled with a gimmel?
A: My *very* question!

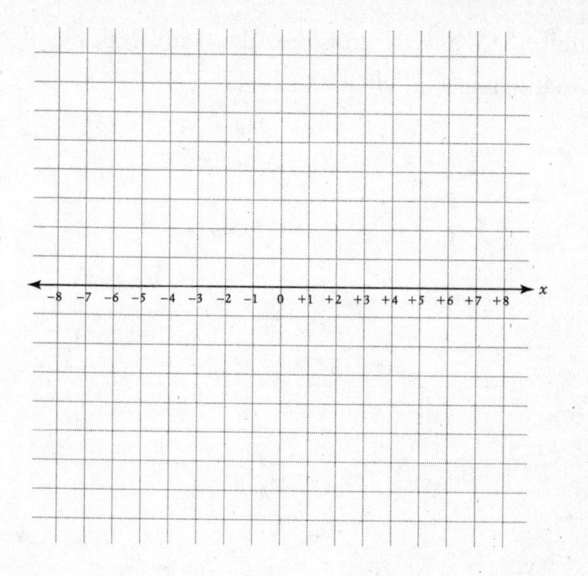

Eventually we are going to think of each point on this plane as being a complex number. We have just identified the points on the horizontal axis with real numbers. Our new imaginary numbers will be positioned elsewhere on the plane. Since we want to envision multiplication by $\sqrt{-1}$ as rotation by 90 degrees, it would seem that we have little choice about placement, and the numbers that are multiples of real numbers by $\sqrt{-1}$ must line up along the vertical axis. But we do have one small choice: should we rotate clockwise or counterclockwise? You can make whichever of these two choices you wish; it isn't going to matter as long as you remain consistent. In the picture that follows I have rotated counterclockwise, so I will stick with this choice.* We will consider the implications of this "small choice" in chapter 12.

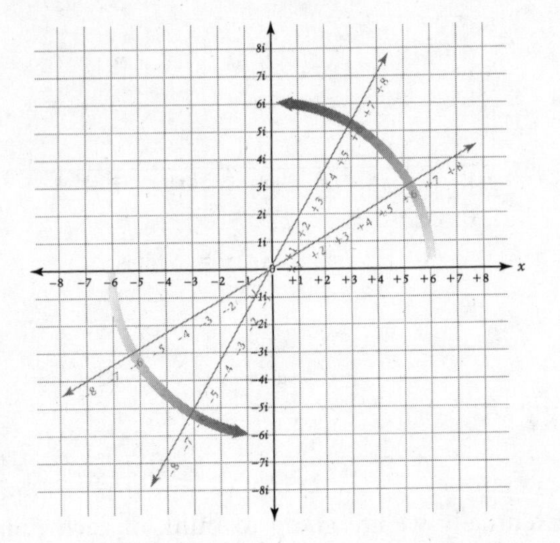

What have we done so far? We have identified the points on the horizontal axis of the Euclidean plane with real numbers, and the points on the vertical axis with numbers that are equal to $\sqrt{-1}$ times real numbers. In Cartesian coordinates, then, a point P of the plane with coordinates $(a,0)$—that is, a point on the horizontal axis—is identified with the real number a. And a point Q of the plane with coordinates $(0,b)$, where b is some real number—that is, a point on the vertical axis—is identified with the imaginary number $\sqrt{-1} \cdot b$, which we can, and will, also write as bi.

*This choice is the "standard" one.

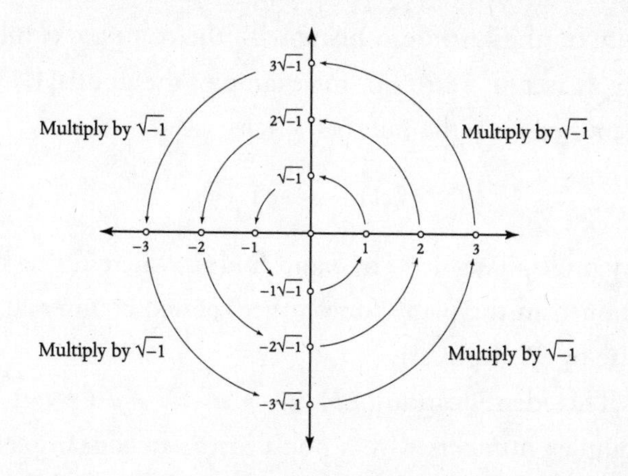

But, of course, we have other points on the Euclidean plane besides those on the horizontal and vertical axes. Can we identify *all* the points on the plane with numbers so as to understand the geometry of multiplication by i as simply rotating the plane by 90 degrees?

For example, consider the point (1,1):

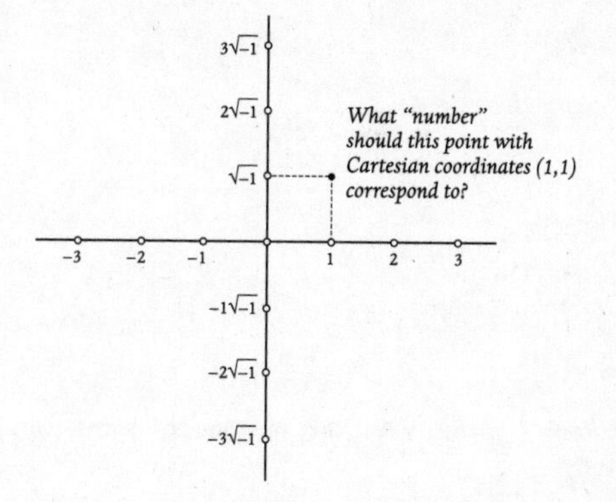

Finding the answer to the question in the diagram is, happily, straightforward. We can identify the point

(a,b) on the Euclidean plane with the complex number $a + bi$ (see p. 146). So, for example, the point (1,1) is identified with the number $1 + i$:

$$(1,1) \longleftrightarrow 1 + i.$$

Try multiplying $1 + i$ by i and finding where it gets positioned in the plane (does it get rotated counterclockwise by 90 degrees?)

This identification $(a,b) \longleftrightarrow a + bi = a + b\sqrt{-1}$, of complex numbers with a point in the Euclidean plane, produces for us a whole *plane of numbers*, extending the old number line. The *real numbers* (the old number line) comprise the horizontal axis; the *imaginary numbers* comprise the vertical axis. In recognition that we have made this identification, we can rename the Euclidean plane the *complex plane.*[8]

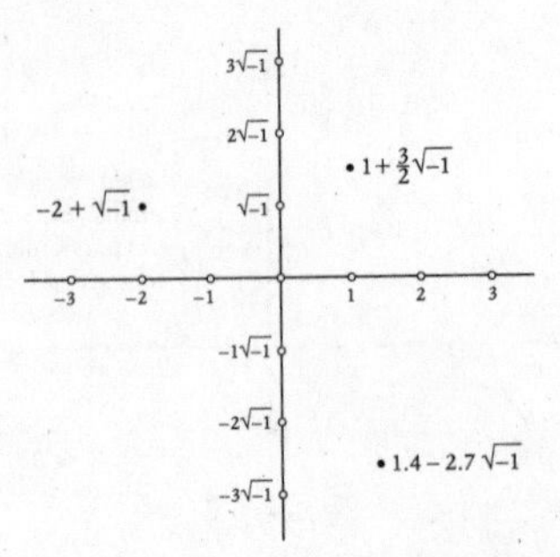

The complex plane, with the position of some complex numbers indicated

50. Thinking silently, out loud.

If we are looking for some internal structure to the act of imagination corresponding to reading "the yellow of the tulip," perhaps we should pay more attention to the word *singing* in Ashbery's

> Singing the way they did, in the old time, we can sometimes see through the tissues and tracings the genetic process has laid down between us and them.

For the one undeniable thing that happens to us when we read is that, in effect, we rarely read silently: we usually hear what we read as spoken by *some* inner voice: we "sing the way they did." As we see, in our mind's eye, the yellow flash upon reading "the yellow of the tulip," we also hear it. We may look at pictures with mute inner voices, or listen to music that way, but some internal speech organ takes over when we read. This is the starting rumination in a Thomas Lux poem: "The Voice You Hear When You Read Silently"

> *is not silent, it is a speaking-*
> *out-loud voice in your head: it is* spoken,
> *a voice is* saying *it*
> *as you read. It's the writer's words,*
> *of course, in a literary sense*
> *his or her* voice, *but the sound*
> *of that voice is the sound of your voice.*
> *Not the sound your friends know*

or the sound of a tape played back
but your voice
caught in the dark cathedral
of your skull, your voice heard
by an internal ear informed by internal abstracts
and what you know by feeling,
having felt . . .

The poem ends this way:

. . . It is your voice
saying, for example, the word barn
that the writer wrote
but the barn you say
is a barn you know or knew. The voice
in your head, speaking as you read,
never says anything neutrally—some people
hated the barn they knew,
some people love the barn they know
so you hear the word loaded
and a sensory constellation
is lit: horse-gnawed stalls,
hayloft, black heat tape wrapping
a water pipe, a slippery
spilled chirr of oats from a split sack,
the bony filthy haunches of cows . . .
and barn is only a noun—no verb
or subject has entered into the sentence yet!
The voice you hear when you read to yourself
is the clearest voice: you speak it
speaking to you.[9]

In contrast to this type of introspection on the internal voices experienced while reading texts, I haven't heard of anyone making as intimate an analysis of what goes on in your head when you read, or re-create, a mathematical argument.

51. The complex plane of numbers.

We can add complex numbers; we can also multiply them. Having called the complex number plane a *plane* gives us some impetus to ask the question: how can we understand these operations, addition and multiplication, geometrically?

The operation of addition has a strikingly elegant, and useful, geometric description in the complex plane. It is called the *parallelogram law.* To find the point $P + Q$ on the complex plane—the point that is the *sum* of P and Q, two complex numbers—you build the (unique) parallelogram in the complex plane that has as three of its vertices the origin of the complex plane, the complex number P, and the complex number Q. Its *fourth vertex* is then the sum $P + Q$:

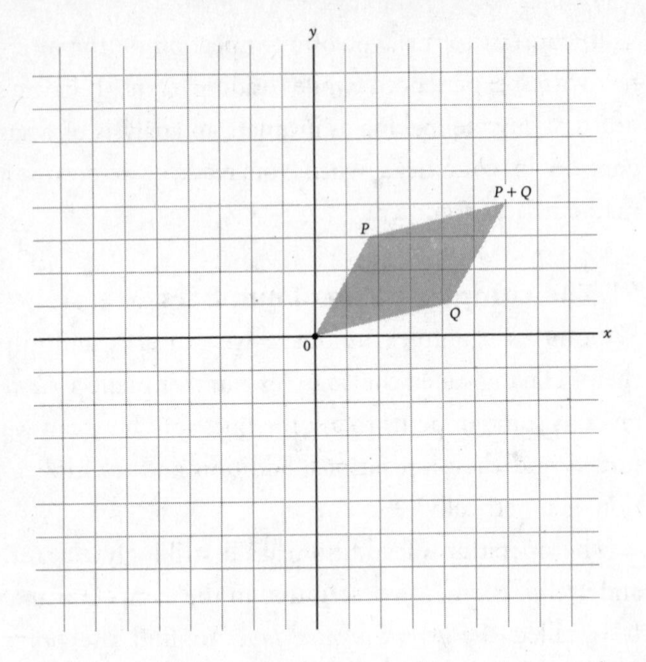

The operation of multiplication of complex numbers has a somewhat more hidden but similarly beautiful and simple geometric description.

We may not be ready for it yet, but an important exercise to do, to begin to "feel" the tight dovetailing of algebra and geometry, is to return to the number $(1 + \sqrt{-3})/2$, "draw" the point on the plane with which $(1 + \sqrt{-3})/2$ has been identified, and figure out the distance of this point from the origin and the angle between the line joining this point to the origin and the horizontal axis. Once you have carefully done this, you might do the same with all three "cube roots of -1" that you have found (in sect. 42) and see where they lie in the plane:

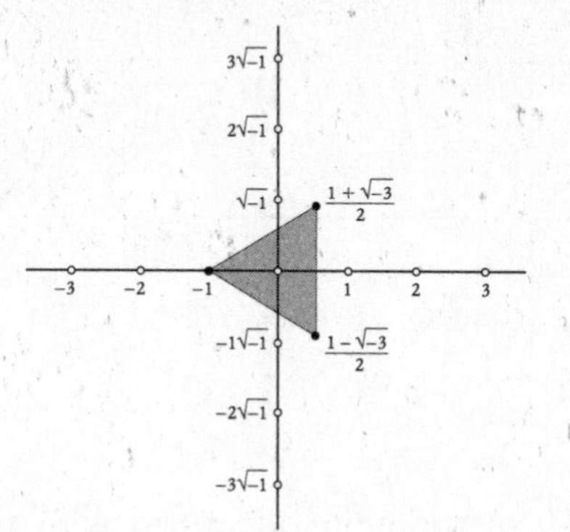

Later, we will want to see what transformation of the plane is made by systematically multiplying each point of the plane by the number $(1 + \sqrt{-3})/2$. But we are perfectly ready now to revisit the geometric meaning of multiplication (of each point in the entire plane of complex numbers) by some real number N.

Recall, for example, that if N is a positive number, then this transformation (multiplication by N) on the old number line is simply a "change of scale," which stretches the line by a factor of N, if N were greater than 1, and compresses it by a factor of N, if N were less than 1 (and does nothing at all if N is 1).

Using a bit of Euclidean geometry, can you check that multiplication by a positive number N on the plane (i.e., the transformation that sends any given point on the plane (a,b), viewed as the complex num-

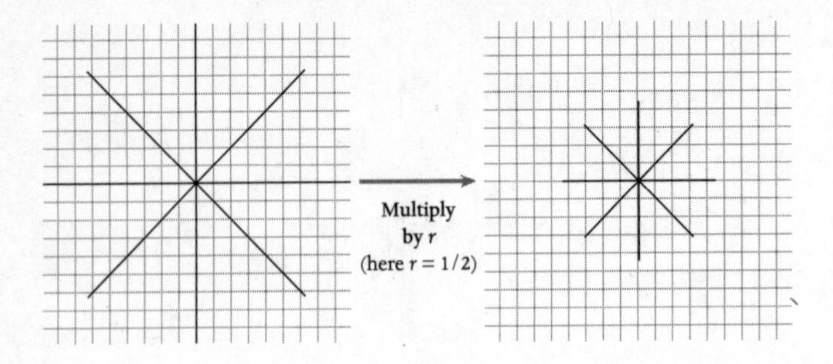

ber $a + bi$, to the point identified with the complex number $N{\cdot}a + N{\cdot}bi$) just stretches or compresses the rays of the plane through the origin by a factor of N?

What is the geometric (motion) picture of the transformation of the plane "multiplication by N" when N is a negative real number—say, when $N = -1$? This, by the way, is an important thing for you to try to visualize in preparation for what we will be doing later on.

52. Telling a straight story.

A friend of mine, a mathematician, after reading an early draft of this book, said that he didn't like my having included so many anecdotes, for example the tales of the contest between Tartaglia and Fiore and of the controversy between Tartaglia and Cardano, in my discussion of $\sqrt{-1}$. "Instead of concentrating on the essential story, the emergent understanding of number," he said, "you have distracted yourself, and confused your text, by describing irrelevant Renaissance pageants."

We were walking on a beach at low tide. Flats of

sand still caught some glints of sun. The shoreline was lost in a gray mist and a gray sea. We could just see figures standing in shallow water, gazing out to sea or digging for clams. We walked toward the water.

"If it is the specific imagination belonging to mathematics that you are trying to capture, then focus on it." On saying this, he interrupted his stride, dropped into a deep-knee-bend position to surprise and catch a sand-colored crab. He held it, a teacup with moveable parts, between two fingers.

"But the basic story is an internal one, logical in the extreme," he said, "and none of the decorations in your essay further the plot." The crab pulsed its claws, fidgeting to flee and bite alternately. My friend moved the animal from his right hand to his left. "Could it be," he asked after a pause, "that the reason you resort to anecdotes is that, in the end, they and their distractions are all we have? Could it be that there is *no* satisfactory way to express the felt experience of the imagination?" The crab jolted again, my friend was caught unawares, and the creature catapulted to the wet sand, paused, then sprinted across the perforated tracks left by the sandpipers pecking for isopods. The crab disappeared into the wash of the waves.

"Even the emphasis, in your essay, on the evolution of notation is off the point," said my friend, "like talking about the type fonts and punctuation in the first edition of Shakespeare's *Sonnets* and thinking that you are talking about the sonnets themselves."

10

SEEING THE GEOMETRY IN THE NUMBERS

53. Critical moments in the story of discovery.

Certain revolutions of thought, we think, announce themselves upon arrival. "We are important," they say. "Take note of us." As for others, their birth is not heralded; their baptismal records are barely legible; and if those records are read at all, they are read only long afterward.

Even when we have palpable historical evidence, it may be hard to weigh the importance of that evidence correctly, for the most minute shifts of thought—a change in notation, the appearance or disappearance of what would seem to be a harmless metaphor—may signal the evolutionary beginning of a new species of idea.

Take inventions. The implications of the invention of the Xerox machine or the personal computer were hardly grasped immediately. How much more difficult would it have been to have understood, at the moment of the invention of *writing* (if there really was such a single moment), what consequences would follow. The fuller reaches and uses of writing could hardly be envisioned before there were whole libraries of written works. This is part of the fun of John Ashbery's account of writing in his "Whatever It Is, Wherever You Are," where he has its inventors already reflecting upon the furthest implications of their happy creation.

Among ideas well documented in our intellectual histories there are few so consequential in radical change of viewpoint as to have construed the sun as fixed, the earth as moving, when no one had done so before. No upheaval of viewpoint has been more immediately seized upon as important.[1] It would be good to understand what these Copernican ideas meant, viscerally, to the Italian inventors of algebra. Did the mathematicians, for example, feel a kinship to Copernicus in the revolutionary nature of their own work? (The second edition of *Ars Magna* was dedicated to Andreas Osiander, the anonymous author of the preface to Copernicus's *De Revolutionibus Orbium*.)

54. What are we doing when we identify one thing with another?

You may have been confused, or perhaps wary, as I have gone about identifying things with other things. For example, I insisted outrageously (in sect. 37) that we *iden-*

tify the real number N with the transformation of the number line consisting of multiplying every number on the line by N. As if this were not enough, we also have *identified* each point (a,b) on the plane with a complex number $a + b\sqrt{-1}$. And now we are right in the middle of the activity of *identifying* complex numbers with transformations of the entire plane. What is this business of *identifying* one thing (in our imagination) with another (which is markedly different from the first thing)? Why do mathematicians do it so much?

The simple answer is that whenever we make any such identification we are signaling a shift, however slight, in point of view. Quite often in teaching math one finds oneself *identifying* things, often by saying that one mathematical thing, X let us call it, *is nothing other than* a different mathematical thing, Y.

> To stipulate a point P in the plane you are doing *nothing other than* stipulating a pair of real numbers a and b that are the Cartesian coordinates (a,b) of the point P.

Unassailable and, better yet, true. Nevertheless, the *nothing other than* in the above statement crosses, with very little fanfare, the bridge between *geometry* and *algebra* that took the centuries between Nicolas of Oresme and Descartes to build.

It is a useful, perfectly benign strategy in the teaching of math to utter such statements as "X is nothing other than *Y*": these statements are often thoroughly graspable, often logically immediate, and yet often effect some needed change of perspective.

55. Song and story.

Ashbery's "the yellow of the tulip" and its accompanying reflections on the invention of writing occur within a literary form quite appropriate for ruminations about writing. The *prose poem* is, by its very label, a hybrid. If a prose writer and a poet were on different sides of a mountain and each felt inspired to burrow a tunnel to join the writer on the other side, the celebratory song they would sing as they met up in the middle would be the prose poem. Here is the poet Baudelaire on the matter:

> Which of us, on our ambitious days, has not dreamed of the miracle of a poetic prose, musical without rhythm and without rhyme, supple enough and abrupt enough to adapt itself to the lyrical movements of the soul, the undulations of revery, the sudden springs of conscience?[2]

And, on the prose writer's side of the mountain, here is the novelist Virginia Woolf (commenting on the essayist Thomas De Quincey):

> But in what form was he to express this, that was the most real part of his own existence? There was none

> ready to his hand. He invented, as he claimed, "modes of impassioned prose." With immense elaboration and art he formed a style in which to express these "visionary scenes derived from the world of dreams." For such prose there were no precedents, he believed; and he begged the reader to remember "the perilous difficulty" of an attempt where "a single false note, a single word in the wrong key, ruins the whole music."[3]

But one needn't look to prose poetry to find traces of hybridization. Poetry itself is a hybrid, as is song, which is the grafting together of words and melody. Only unlike song, poetry's melody emanates from the words themselves, and words seem to emanate from its melody. If poetry, then, is like song, prose poetry is like talking blues.

56. Multiplying in the complex plane. The geometry behind multiplication by $\sqrt{-1}$, by $1 + \sqrt{-1}$, and by $(1 + \sqrt{-3})/2$.

The recipe for multiplication of complex numbers has two equivalent descriptions, one *algebraic* and the other *geometric*.

We have seen the algebraic rule, and here is a quick review. To multiply two complex numbers, such as $3 + 4\sqrt{-1}$ and $5 + 6\sqrt{-1}$, you work out four more elementary multiplication problems, multiplying the real and imaginary parts of the first complex number each by the real and imaginary parts of the second, and then adding up all four products, as done in this endnote.[4]

For example, multiply *any* complex number, $a + bi$, by i and you get

$$i \times (a + bi) = ai + bi^2 = -b + ai.$$

Multiplication by i then takes the complex number $a + bi$ (which, if a and b are positive, is situated a units to the right of the imaginary-number axis and b units above the real-number axis) to the complex number $-b + ai$ (which is b units to the left of the imaginary-number axis and a units above the real-number axis). Can you see that this operation is just the transformation of the plane that rotates the entire plane through an angle of 90 degrees about the origin, in the counterclockwise direction?

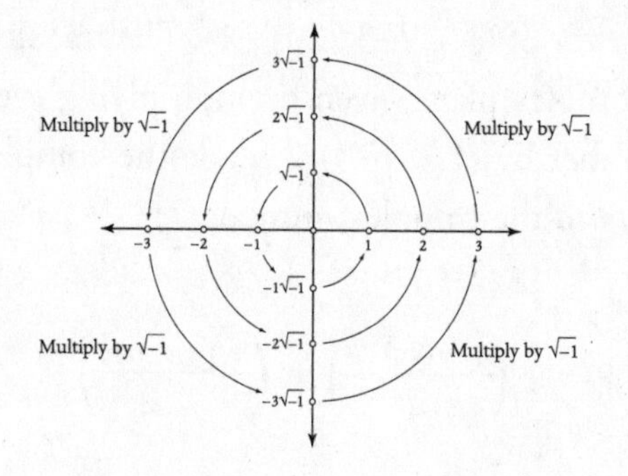

If you have done this multiplication, I have two further exercises for you.

First, try your hand at visualizing the transformation of the complex plane that is given by multiplying every complex number by the complex number $1 + i = 1 + \sqrt{-1}$. Visualize it, and also describe it in words, as we have done with the transformation given by multiplication by i.

A hint (which you may not need): first square the complex number $1 + i$; then visualize what transformation of the complex plane corresponds to this; and then attack the problem.

Our second job here is to return to the complex number

$$\frac{1 + \sqrt{-3}}{2} = \frac{1}{2} + \left(\frac{\sqrt{3}}{2}\right) i.$$

with which we have already done experiments (recall that the *cube* of $(1 + \sqrt{-3})/2$ is -1). The transformation of the complex plane given by multiplying every complex number by $(1 + \sqrt{-3})/2$ sends the complex number $a + bi$ to the complex number

$$\left(\frac{a - \sqrt{3} \cdot b}{2}\right) + \left(\frac{\sqrt{3} \cdot a + b}{2}\right) i$$

But, as in the case of multiplication by i and by $1 + i$, there is a simple geometric description of this transformation of the plane.

Can you guess a simple geometric description of this transformation?

A hint to help in your guess: think of the result you got when you cubed $(1 + \sqrt{-3})/2$.

Now, can you assure yourself that your guesses in both exercises are correct?

To do these exercises is to get a vivid sense of the connection between the underlying algebra of the complex numbers and their underlying geometry. Whenever two different kinds of intuition meet and mesh (in this case, a *geometric* intuition is being overlaid onto an *algebraic* structure), a new, more powerful *hybrid* intuition is the result. And we have work to do to lodge this newcomer firmly in our imaginations.

57. How can I be sure my guesses are right?

If you happen to remember some of your high school algebra and analytic geometry, you are in a position to prove your guesses to be correct (if you have in fact made the correct guesses). To *prove* that something is correct, you must, of course, have an unambiguous understanding of all the concepts involved, an unambiguous understanding of all the operations used, and a clear formulation of all the *truths* you have taken to be self-evident, or at least hypothesized, in your argumentation. The idea of proof is one of the great lessons taught in high school geometry. But let us assume that you have forgotten all your high school mathematics, that you are, mathematically, tabula rasa, that you know no theory whatsoever. You have, nevertheless, gotten to this point in the book, and you want to gauge whether your guesses in the last section are correct.

A simple strategy is: do what you can! The wonderful thing about mathematics is that, in the end as well as in the beginning, it can depend upon no authority other than one's own (your own) mind; its verification comes from thinking alone, an activity open to anyone. If we have no theoretical equipment, we use the mathematical eyes and ears with which we were born and just experiment with our guesses to see whether we have faith in them. Of course, experimentation alone will not be able to convince us with certainty, but experimentation has, nevertheless, essential lessons to teach us.

For example, to approach the question in the heading of this section, take a few specific complex numbers Q and multiply each of them by $(1 + \sqrt{-3})/2$, plot the answers (approximately, if you can't do this exactly), and then see whether your guessed geometric description of the transformation of the complex plane given by multiplication by $(1 + \sqrt{-3})/2$ approximately fits the data that you are accumulating. This type of experimental mathematical work may not lead to a *proof* that your guess is correct, but it has important consequences. For one thing, it will begin to build in you a sense that there may indeed be two ways to describe multiplication of complex numbers, one algebraic and the other geometric, giving us a bridge between algebra and geometry.

When we began the project of understanding numbers, we were led to contemplate objects like $\sqrt{-1}$, to make them seem real to us, but did we suspect that this project would take the turn it did, of building a bridge between two quite different modes of intuition?

Perhaps, if we were supreme algebraists, with algebra as our mother tongue, we wouldn't need such a bridge. A friend of mine, for example, is gifted with a strong algebraic intuition and insists that, despite the mild uneasiness expressed by the early users of imaginary numbers, it is simply wrong to think there was some essential *lacuna of the imagination* regarding these numbers as long as there was no geometric picture of them. "Imaginary numbers behave according to definite and clearly understood rules" he says. "Nothing more is really needed."

The alternative view, however, is that a great achievement of mathematics is to make supple, make ever more fine, and audaciously broaden our intuitions. If ever faced with a choice of intuitions, *algebraic* or *geometric* for example, we would do well to follow the wise advice given in the film *Beat the Devil* on whether to purchase a Rolls-Royce or a Cadillac, and to recognize that we are in a position to have both.

58. What is a number?

We have been talking of the numbers 2, $\frac{1}{2}$, and $\sqrt{-1}$ and various of its companions *as giving rise to the transformation of the complex number plane that results from multiplying all other numbers by them.* Multiply by the number 2, for example, to get the transformation of doubling, or *scaling up the complex plane by a factor of 2*; do the same with the number $\frac{1}{2}$ to get halving, or *scaling down the complex plane by a factor of 2*; do it with the complex number $\sqrt{-1}$ and you get the transformation of *rotating the complex plane counterclockwise, around the origin, by 90 degrees.* What precedents are there for this kind of thinking about numbers?

Readers who remember studying Euclid's *Elements of Geometry* will see an analogy between thinking of numbers as "scaling transformations" and the Euclidean theory of "proportions" (number as *ratio*). Nicolas of Oresme, as we have already discussed, worked with *proportions*, which he called *intensities*. To remind us of the flavor of this approach to number, here is Isaac Newton's explanation of what he means by number (in *Universal Arithmetick*):

> By *Number* we understand, not so much a Multitude of Unities, as the abstracted Ratio of any Quantity, to another Quantity of the same Kind, which we take for Unity.[5]

To view number as transformation, as we have done, is to take a dynamic approach to this older consideration of number as proportion, or ratio; but here we envision the proportion in question as being *animated* by the transformation of the plane that is the appropriate expansion or contraction (and also, in the case of complex numbers, perhaps rotation as well).

59. So, how can we visualize multiplication in the complex plane?

Guess first, before reading this!

Let us take any nonzero complex number, $P = a + bi$, and describe its position in the complex plane. Recalling our discussion of section 23, we have two alternative methods of doing this. We might pinpoint P either by giving its Cartesian coordinates,

$$a + bi \longleftrightarrow (a,b),$$

or by stipulating its polar coordinates. This would require providing its *magnitude* r (which, you will recall, is the distance between the origin, 0, and point P) and its *phase* α. Recall (sect. 23) that the phase of P is the angle between the line segment from 0 to P and the part of the x-axis to the right of 0.

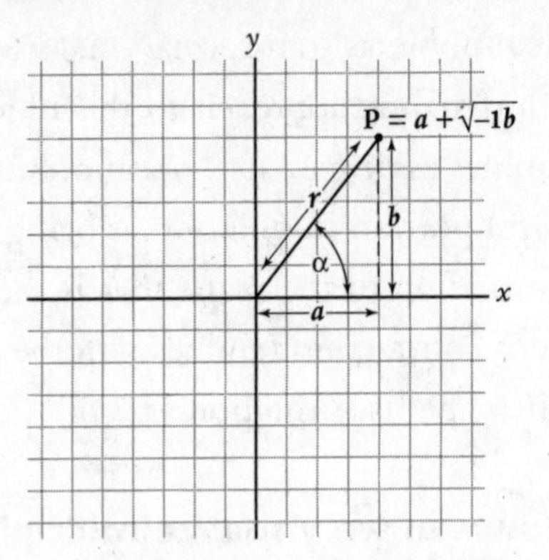

Since we have two "languages" (Cartesian and polar) for describing the position of complex numbers, we have, of course, the problem of translation from one mode to the other.

The key to the geometric description of multiplication by P is in the polar coordinate description of P—that is, the data (r,α). The transformation of the complex plane that sends any complex number Q to the product $P \cdot Q$ can be given by performing two separate transformations, one after the other—the first determined by the *magnitude* r of P and the second by its

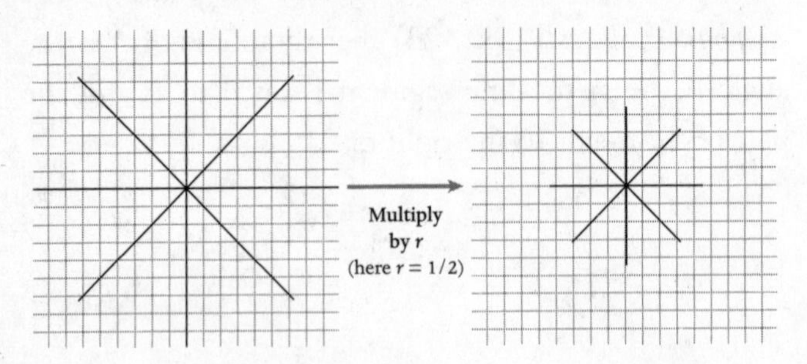

phase α. The first of these transformations rescales the plane by multiplying by the positive real number r (bottom of p. 192); the second rotates the plane counterclockwise about 0 through the angle α:

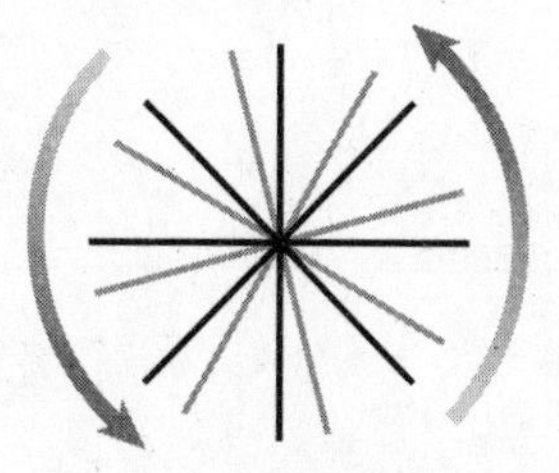

Rotate by angle α

Was that your guess?

Here is one way of summarizing this quite tidy setup. We have two kinds of operations for complex numbers (addition and multiplication) and two ways in which we can describe the complex plane (by Cartesian and polar coordinates). The Cartesian description makes addition "easy," while the polar description makes multiplication "easy." For, as we have seen, in the Cartesian coordinate description of complex numbers, addition is simply given by:

$$\begin{array}{c} a + \sqrt{-1} \cdot b \\ \underline{c + \sqrt{-1} \cdot d} \\ (a + c) + \sqrt{-1} \cdot (b + d). \end{array}$$

As for multiplication, the product of two complex numbers has its magnitude equal to the *product* of the magnitudes of its two factors, and its phase (angle) equal to the *sum* of the phases of its two factors.[6] Here is an example. Consider two complex numbers U and V and their product W:

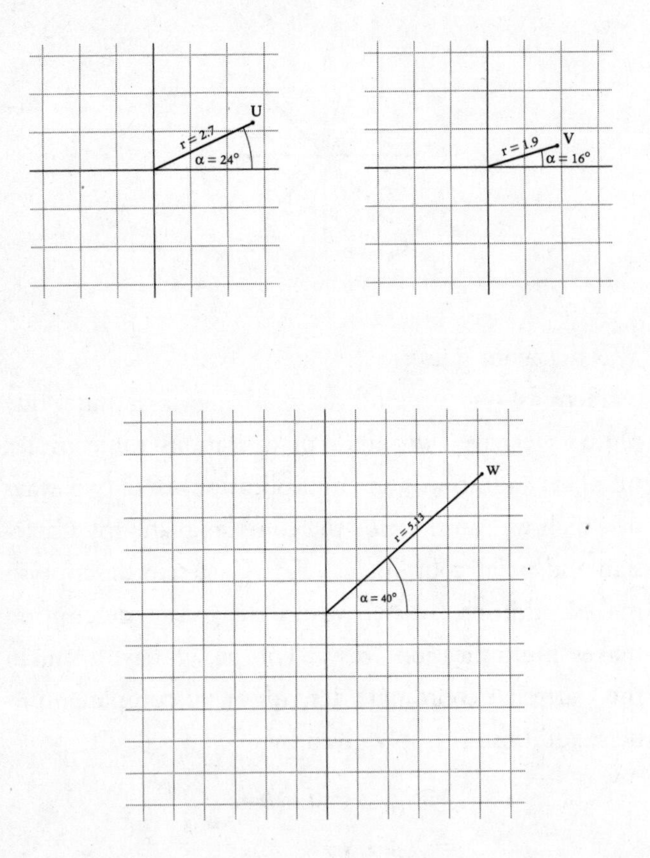

U has magnitude $r = 2.7$ and phase $\alpha = 24°$.
V has magnitude $r = 1.9$ and phase $\alpha = 16°$.

So if $W = U \cdot V$, then W has magnitude equal to the product of the magnitudes of U and V,

$$r = 2.7 \times 1.9 = 5.13,$$

and phase equal to the sum of the phases of U and V,

$$\alpha = 24° + 16° = 40°.$$

You are now ready for another exercise. Given a nonzero complex number P with magnitude r and phase α, can you describe the magnitudes and phases of the two complex numbers that are *square roots* of P (yes, there are two) and the magnitudes and phases of the three complex numbers that are *cube roots* of P (yes, there are three)?

PART III

11

THE LITERATURE OF DISCOVERY OF GEOMETRY IN NUMBERS

60. "These equations are of the same form as the equations for cosines, though they are things of quite a different nature."

When was the geometric view of complex numbers discovered? Although there were earlier hints of the link between the solution of polynomial equations and trigonometry,[1] we know that Abraham De Moivre, by 1707, had perceived the analogy between the geometric problem of cutting an arc of a circle into n equal parts and taking the nth root of a complex number. To explain De Moivre's insight, let us take the case of $n = 2$ and consider the reverse operation of doubling an angle and squaring a complex number. We must do, or at least cite, a bit of geometry here.

Consider a pie wedge in a circle of radius 1,

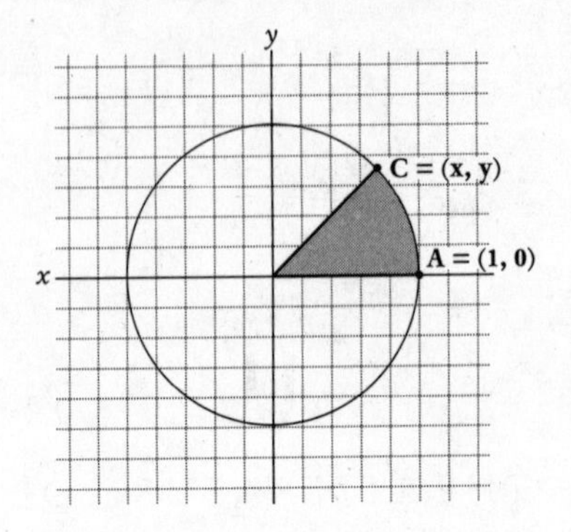

and cut the wedge into two equal portions,

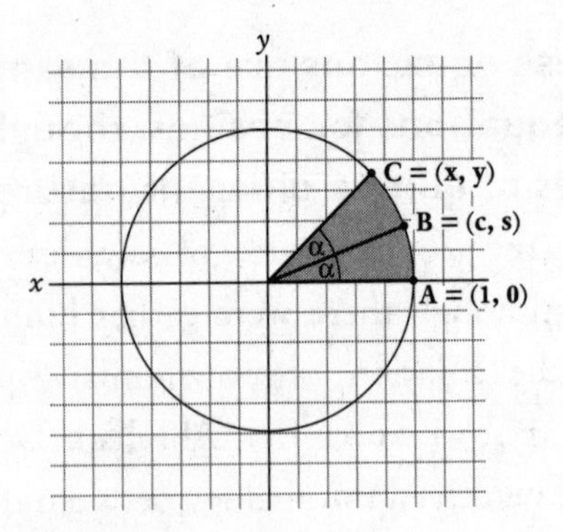

Now consider the coordinates of the three points labeled *A*, *B*, and *C* in this diagram. Point *A* lies on the horizontal line through the origin, is of distance 1 from the origin, and is to the right of it, so it has coordinates (1,0). Suppose point *B* has coordinates (c,s) for particular numbers c and s (these are the "cosine" and the "sine," respectively, of the angle α of each pie wedge, but you needn't consider this for the purposes of our discussion here). One can now prove (although we will not do it) that if (x,y) denotes the coordinates of point C, we may express (x,y) in terms of (c,s), the coordinates of point B, by the equations

$$x = c^2 - s^2 \text{ and } y = 2cs.$$

What De Moivre noticed was that one gets exactly the same expressions, $c^2 - s^2$ and $2cs$, if one squares the complex number $c + \sqrt{-1}\cdot s$. That is,[2]

$$(c + \sqrt{-1}\cdot s)^2 = (c^2 - s^2) + \sqrt{-1}\cdot(2cs).$$

In other words,

$$(c + \sqrt{-1}\cdot s)^2 = x + \sqrt{-1}\cdot y.$$

In effect, when you *square* a complex number, you *double* its phase. Going further, when you *cube* a complex number, you *triple* its phase, and so on. De Moivre noticed, and expressed, the general pattern as a similarity of formulas of "things of quite a different nature."

Now, a similarity of formulas in two different mathematical contexts immediately suggests the possibility of a deeper structural analogy relating these contexts.

Here is how this analogy appears in De Moivre's 1738 article "Of the Reduction of Radicals to More Simple Terms" (the emphasis is mine):[3]

> By putting $2x$ equal to the n-th root of $(a + \sqrt{b})$ plus the n-th root of $(a - \sqrt{b})$ and making the n-th root of $(a^2 - b)$ equal to m, then expounding n succes-

sively by 1, 2, 3, 4, 5, 6, &c. there will arise the following equations:

1st. $x = a$
2d. $2x^2 - m = a$
3d. $4x^3 - 3mx = a$
4th. $8x^4 - 8mx^2 + m^2 = a$
5th. $16x^5 - 20mx^3 + 5m^2x = a$
6th. $32x^6 - 48mx^4 + 18m^2x^2 - m^3 = a$
7th. $64x^7 - 112mx^5 + 56m^2x^3 - 7m^3x = a$
&c.

Now these equations are of the same form as the equations for cosines, though they are things of quite a different nature. Thus, let r be the radius of a circle, c the cosine of any given arc, and x the cosine of another arc, which is to the former, as 1 to n. Then it will be:

1st. $x = c$
2d. $2x^2 - r = c$
3d. $4x^3 - 3rx = c$
4th. $8x^4 - 8rx^2 + r^2 = c$
5th. $16x^5 - 20rx^3 + 5r^2x = c$
6th. $32x^6 - 48rx^4 + 18r^2x^2 - r^3 = c$
7th. $64x^7 - 112rx^5 + 56r^2x^3 - 7r^3x = c$
&c.

What De Moivre is observing as a similarity between sets of equations governing *things of quite a different nature* is now packaged as an equation (referred to as "De Moivre's formula"):

$$(\cos \theta + \sqrt{-1} \sin \theta)^n = (\cos n\theta + \sqrt{-1} \sin n\theta).$$

This formula seems to "say it all," but by its very compactness it fails to express the spirit in which De Moivre viewed his finding that his equations are of the same form as the *equations for cosines.* De Moivre then uses the similarity between his two collections of formulas to fashion a connection between extraction of *n*th roots and the cutting of a wedge (see figures on p. 200) into n equal pieces. Most important for our particular story, and for our understanding of Dal Ferro's formula, is the special case of when $n = 3$: the connection between the extraction of cube roots and angle trisection. But De Moivre sets himself the general problem "to extract the *n*-th root of the impossible binomial $a + \sqrt{-b}$"—which he solves by cutting the corresponding wedge into n equal pieces, or, as he puts it more succinctly, with a "table of sines." He finds, for example, that the three cube roots of $81 + \sqrt{-2700}$ are

$$\frac{9}{2} + \left(\frac{1}{2}\right)\sqrt{-3},\quad \frac{3}{2} - \left(\frac{5}{2}\right)\sqrt{-3},\ \text{and}\ -3 + 2\sqrt{-3},$$

and then writes:

> There have been several authors, and among them Dr. Wallis, who have thought that those cubic equations, which are referred to the circle, may be solved by the extraction of the cube root of an imaginary quantity, as of $81 + \sqrt{-2700}$, without any regard to the table of sines: but that is a mere fiction; and a begging of the question; for on attempting it, the result always recurs back again to the same equation as

> that first proposed. And the thing cannot be done directly, without the help of the table of sines, specially when the roots are irrational; as has been observed by many others.[4]

This paragraph, written close to two centuries after Bombelli wrote his *L'Algebra*, furthers the ideas of Viète and Girard mentioned previously and addresses what we have called "Bombelli's puzzle" (i.e., how to interpret Dal Ferro's formula when the "indicator" is negative, as discussed in section 33). It is also here that geometry has begun to clarify the algebra of complex numbers. The Dr. Wallis referred to is the English mathematician John Wallis, who in his 1673 text *Algebra* expresses imaginary quantities as "mean proportions" between positive quantities and negative ones. Wallis drives home the analogy between this and geometric constructions: "What hath been already said of $\sqrt{-bc}$ in algebra (as Mean Proportional between a Positive and a Negative Quantity:) may be thus Exemplified in Geometry."[5]

This sentence, which is a prelude to a geometric construction in Wallis's text that places $\sqrt{-bc}$ in the plane but out of the number line contains the essence of the idea that will fully germinate over a century later.

Let us return, for a moment, to this small difference between the nature of the equation we now call "De Moivre's," which ties complex numbers to trigonometry, and the manner in which De Moivre expressed himself (as quoted in the heading of this section). De Moivre's language focuses on the *analogy* between complex numbers and trigonometry; the formula that bears his name effects an *identity*.

Analogy is everywhere in mathematics; when it is most fruitful it is most unstable, goading mathematicians into producing larger structures in which the *analogy* becomes an *equality* (these "larger structures" are ripe, to be sure, with new analogies). Having said this, perhaps I should, in fairness, quote the great twentieth-century mathematician André Weil, whose sentiments regarding the evolution of *analogy* in mathematics have a more somber tint. Here are his thoughts on the matter, the music of which has mesmerized me for decades, even as the thought it expresses has appalled me:

> Nothing is more fruitful—all mathematicians know it—than those obscure analogies, those disturbing reflections of one theory on another; those furtive caresses, those inexplicable discords; nothing also gives more pleasure to the researcher.
>
> The day comes when this illusion dissolves: the presentiment turns into certainty; the yoked theories reveal their common source before disappearing. As the *Gita* teaches, one achieves knowledge and indifference at the same time. Metaphysics has become mathematics, ready to form the material of some treatise whose cold beauty has lost the power to move us.[6]

Oh, but how grim education would be if knowledge and indifference were ineluctably conjoined; how inexplicable the shouts of "Eureka!" of discoverers through the ages; how bizarre the tale of Pythagoreans murdering the one who revealed that the square root of 2 is not expressible as a fraction!

Like Pythagoras, Abraham De Moivre has been the subject of a number of biographies, and as they did for his great predecessor, some of these biographies have made quite fanciful statements about their subject. For example, it is often related that De Moivre calculated that each day he slept fifteen minutes more than the previous day, and based on this calculation he correctly predicted the exact day of his death (i.e., when the total reached twenty-four hours).[7]

In the eighteenth century, after De Moivre's work, the application of sines and cosines was ubiquitous in calculations involving complex numbers, and consequently geometric issues were present in some form. For example, it is difficult for a modern reader of Leonhard Euler's 1749 article "Recherches sur les racines imaginaires des équations"[8] not to think of the complex plane as relevant to Euler's calculations. Nevertheless, one finds no hint at that time of any explicit representation of the set of complex numbers as points on a plane.

61. A few remarks on the literature of discovery and the literature of use.

De Moivre's writings vividly establish a yoking between geometry and the algebra of the complex numbers. *Yoking* may be an appropriate word, with its implication that we have two separate but somehow connected intuitions: the geometry and the algebra ranged side by side, each carrying within it the seeds of the structure of the other.

When the fully integrated geometric-algebraic viewpoint finally emerged explicitly in print, it came from many sources, but not from the great mathematicians of the time.[9] It may be that for the central figures in mathematics of the eighteenth and early nineteenth centuries—Leonhard Euler (1707–83), for example, or Carl Friedrich Gauss (1777–1855)—De Moivre's yoking was more than sufficient. Perhaps the strength of their mathematical perception was such that no further concrete imagining was necessary for them to do their work, or perhaps none was lacking. They were just busy discovering wonderful mathematics.

Gauss, for example, proved in his dissertation (defended in 1799) that any nonconstant polynomial with real coefficients is the product of factors that are either linear or quadratic polynomials with real coefficients, and therefore any such polynomial has a complex (possibly real) root. This is a version of what we now know as the *fundamental theorem of algebra.*[10] He did this with the De Moivre tools at his disposal, with hardly a mention of complex numbers, let alone geometry.

The first account of the geometry of complex numbers was given by the Danish-Norwegian Caspar Wessel, a professional surveyor, in a paper originally presented to the Royal Danish Academy of the Sciences in Copenhagen in 1797 and published in 1799 in the *Mémoires* of that academy. This paper, apparently, attracted few readers—until a French translation was published a century later ("La représentation analytique de la direction"). The second account, presenting roughly the same viewpoint, was published in Paris in 1806 by the Swiss bookseller and amateur mathematician Jean-Robert Argand ("Essai sur une manière de représenter les quantités imaginaires dans les constructions géométriques"); this, likewise, seems to have been little read.

A similar treatise was published a few years later by J.-F. Français ("Nouveaux principes de géométrie de position, et interprétation géométrique des symboles imaginaires").[11] In his article, Français disclaims any originality for the basic idea, saying that he found among the papers left by his deceased brother a letter from the mathematician Adrien-Marie Legendre, in which the geometric representation of complex numbers was briefly described. Français reports that Legendre considered this entire issue (representing complex numbers geometrically) as something of a curiosity (*"comme objet de pure curiosité"*), which someone had told him about. Legendre didn't say from whom he learned it, and there seems to be no further mention of this matter in the rest of Legendre's writings. Français goes on to express the wish that whoever initially had the idea of geometrizing complex numbers (*"le premier auteur de ces idées"*) would make himself and his theory known.

Two months later Argand came forward to claim the credit, announcing in a note that Legendre had, in fact, first learned of these ideas from him, and saying that Français's wish had long been answered by his, Argand's, 1806 publication.[12] Argand goes on to give a further exposition of his theory, together with an (incorrect) simplification of an (incorrect) proof proposed by d'Alembert of the fundamental theorem of algebra.[13] Argand also claims in this note that

$$\sqrt{-1}^{\sqrt{-1}}$$

cannot be expressed as a complex number (which is not true: this concoction can be given a natural enough interpretation, which has a real-number value, as already seen by Euler).

Both Français's article and Argand's response were published in a corner, labeled "Philosophie Mathématique," of the mathematics journal *Annales de mathématiques pures et appliquées*, edited by Joseph-Diez Gergonne. Gergonne played the role of editor-participant, quite enthusiastic about the geometric ideas on complex numbers (the catchphrase for these ideas is *"géométrie de position,"* which sounds somewhat redundant to me). Gergonne would get his friends to submit articles and letters on the subject;[14] he would appeal to *X* to respond to the article by *Y*, and publish article and rebuttal in the same issue of his journal; he would pepper the articles with his own footnotes, expressing his enthusiasms, explaining things more fully, objecting to viewpoints presented, or playing spry referee, mediating between opposing views. For example, toward the end of Français's submission to the *Annales*, Français writes:

> Such is the sketch, very much abbreviated, of the new principles on which it is convenient, and even necessary, to base the theory of "Geometry of Position" which I submit to the judgment of geometers. Since these principles are in formal opposition to current ideas about the nature of imaginary quantities, I expect numerous objections; but I dare to believe that a deep examination of these same principles will find them correct, and the consequences that I deduce from them, no matter how strange they may first appear to be, will nevertheless be judged to conform to the most rigorous rules of dialectic.[15]

Gergonne's page-long footnote at the end of this article takes issue with the question of whether Français's ideas are all that strange. "Far be it from me," says Gergonne, "to deflate M. Français. But I want to show that these ideas are not at all so strange as not to be capable of germinating in several heads at once." Gergonne then goes on to report on a footnote he had written for an article published two years earlier in the *Annales*, where he himself, Gergonne, found it reasonable to clarify the presentation of that article by rearranging a series of complex numbers in a rectangular array similar to their geometric placement in Français's theory.

Gergonne asked J.-F. Servois to comment on the articles published by Français and Argand. Servois responded with a letter to the editor duly tabulating Argand's errors and noting that much of what Français asserted was not substantiated. Servois then went on to offer his doubts about the whole geometrizing enterprise, referring to it as a "mere analogy," and elsewhere as "peculiar notation." He writes:

> In setting up the foundations of an extraordinary doctrine, somewhat opposed to received principles [principes reçus], in a science such as mathematical analysis, mere analogy is hardly a sufficient mode of reasoning.[16]

Servois writes that he doesn't see Français's or Argand's theory as anything more than a "geometric mask" pasted onto analytic forms, and that the algebra, unadulterated by geometry, is simpler and more efficacious. Gergonne, to be sure, has been interrupting Servois's text all along with footnotes, and at this point Gergonne writes:

> Does Monsieur Servois consider it unimportant to see, at last, algebraic analysis stripped of its unintelligible and mysterious forms, of the nonsense that limits it and makes of it, so to speak, a cabalistic science?[17]

All three treatments, Wessel's, Argand's, and Français's, construe $\sqrt{-1}$ as a kind of "geometric mean" between +1 and −1, and by that logic they position it symmetrically between +1 and −1 in the plane (as we have done). As we have seen, John Wallis, more than a century earlier, also viewed square roots of negative quantities as geometric means. It is in Argand's memoir that one finds the role of *number as transformation* explained; and, by that logic, if multiplication by −1 is rotation by 180 degrees, then −1, as transformation, has a clear square root (in fact, two of them).[18]

There is an arresting quaintness to how Français, in his treatise, chooses to encapsulate his discovery of the way to imagine imaginary numbers. He formulates his punch lines as *Theorems* and *Corollaries*. Here is one of them:

> *Corollary 3. Imaginary quantities* are just as real as positive numbers and negative numbers, and only

> differ in their position, which is perpendicular to the latter.[19]

Here one sees Français struggling to package his insight into a logical syntax that isn't exactly appropriate to it (as Servois points out). But its importance is not quite in its *logical structure*—rather, Français's "corollary" is the written record that a significant leap of the imagination has been made.

This lively interchange, with Français and Argand (and Gergonne on the sidelines) promoting the geometric viewpoint, and Servois wondering whether geometrization is but a *mask* to obscure algebra, occurred in 1813. How different such a conversation might have been one year later, when Augustin-Louis Cauchy's first memoir on contour integrals in the complex plane was published, ushering in the immense "literature of use" of the geometric viewpoint. For in Cauchy's article, the profound geometry of the plane is brought into play as a vital and inseparable aspect of the study of complex numbers, just as the algebra of complex numbers clarifies the geometry. And Cauchy's article is, as most advances are, just the beginning.

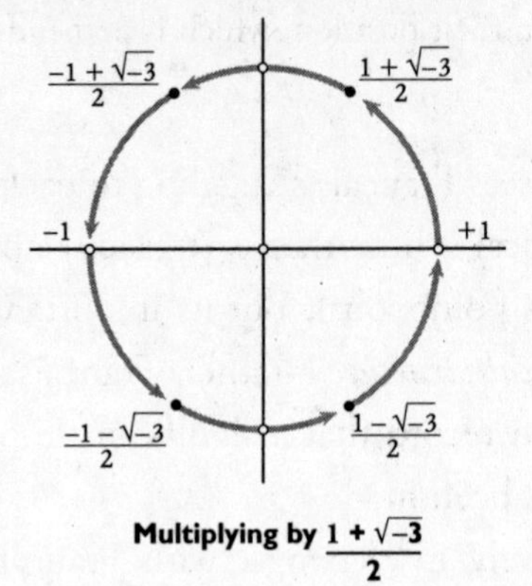

Multiplying by $\frac{1+\sqrt{-3}}{2}$

12

UNDERSTANDING ALGEBRA VIA GEOMETRY

62. Twins.

It is time to take up an element of "choice" in our identification of complex numbers with points in the Euclidean plane:

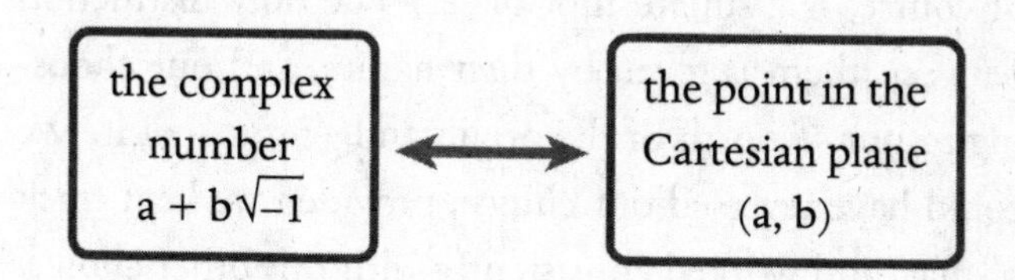

The number $i = \sqrt{-1}$ is identified with the point whose coordinates are (0,1); that is, with the point 1 unit north of the origin. Recall that we chose to view "multiplication by $\sqrt{-1}$" as a rotation by 90 degrees *counterclockwise* about the origin.

A good test of whether we have understood this passage from complex numbers to points on the plane is to ask ourselves what would be different if we had, perhaps perversely, pictured $\sqrt{-1}$ as identified with the point in the plane 1 unit *south* of the origin, rather than north? That is, what if we had identified "multiplication by $\sqrt{-1}$" as rotation of the plane 90 degrees *clockwise*; or, what would be the same, 270 degrees counterclockwise?

The brief answer here is that nothing whatsoever would change, except for the curious fact that i would be playing the role that $-i$ plays in our identification, $-i$ would be playing the role that i plays, and, more generally, the complex number $a - bi$ would be playing the role that $a + bi$ plays. Behind this lies a surprise, and a curious mirror. There is no intrinsic (algebraic) way of distinguishing $+\sqrt{-1}$ from $-\sqrt{-1}$. Each of them, of course, is a square root of -1. The only distinction between them is given by their names, and our choosing to put $+i$ north of the origin and $-i$ south of it. We could have reversed our choice, provided we kept track of that, and worked consistently with this other choice. These entities, $+i$ and $-i$, are twins, and the only breaking of their symmetry comes from the way in which we *name* them. Imagine the analogous moment when parents of newly born identical twins choose names for their two children, thereby making the first, and immensely important, distinction between them.[1]

A reader of an early draft of this book asked why I called $+i$ and $-i$ twins, but did not, for example, call $+1$

and -1 twins. Here is why. The number $+1$ is distinguished from the number -1 by some purely algebraic property (for example, it is equal to its square), while there is no analogous property, described entirely in terms of addition and multiplication, that distinguishes $+i$ from $-i$ and yet does not, either directly or indirectly, make use of the names we gave to them.

The complex numbers $a + bi$ and $a - bi$ are called *conjugates*. And the act of "conjugation"—that is, passing from a complex number to its conjugate (or equivalently, reversing the sign of the imaginary part of a complex number)—is a basic *symmetry* of the complex-number system. There are other number systems that admit a bewildering collection of symmetries, of internal mirrors. One of the great challenges to modern algebra is to understand, and use, these internal mirrors.

63. Bombelli's cubic radicals revisited: Dal Ferro's expression as algorithm.

If $P = a + bi$ is a complex number, then, as we have just discussed, $a - bi$, its *conjugate*, is something of an algebraic twin of P. The sum of any complex number P and its conjugate is twice the real part of P; that is,

$$\begin{array}{c} a + \sqrt{-1}\cdot b \\ \underline{a - \sqrt{-1}\cdot b} \\ 2a + \sqrt{-1}\cdot 0 = 2a \end{array}$$

and the sum is therefore a real number.

Let us now return to Dal Ferro's expression that Bombelli contemplated so long ago,

$$X = \sqrt[3]{\frac{c}{2} + \sqrt{\frac{c^2}{4} - \frac{b^3}{27}}} + \sqrt[3]{\frac{c}{2} - \sqrt{\frac{c^2}{4} - \frac{b^3}{27}}},$$

which purported to be the solution to the general cubic equation $X^3 = bX + c$, and which, as we have seen, has a somewhat alien aspect when the expression under the square roots is a negative real number. It is in this circumstance that our cubic equation will have three real-number solutions. Suppose, then, that $c^2/4 - b^3/27$ is negative, in which case

$$\frac{c}{2} + \sqrt{\frac{c^2}{4} - \frac{b^3}{27}}$$

and

$$\frac{c}{2} - \sqrt{\frac{c^2}{4} - \frac{b^3}{27}}$$

are conjugate complex numbers. For short, call these conjugate complex numbers D and $\bar{\mathrm{D}}$. That is,

$$D = \frac{c}{2} + \sqrt{\frac{c^2}{4} - \frac{b^3}{27}} \quad \text{and} \quad \bar{D} = \frac{c}{2} - \sqrt{\frac{c^2}{4} - \frac{b^3}{27}}.$$

Our solution X then has the "shortened" expression,

$$X = \sqrt[3]{D} + \sqrt[3]{\bar{D}}.$$

Putting together what we have discussed thus far, we can control, somewhat, the ambiguity in Dal Ferro's puzzling expression (the ambiguity we signaled in section 43) if we interpret the cube roots $\sqrt[3]{D}$ and $\sqrt[3]{\bar{D}}$, of the conjugate complex numbers D and $\bar{D}$, as *conjugate* complex numbers as well. Why might we be motivated to do this?

We know that any (nonzero) complex number has three cube roots, so $\sqrt[3]{D}$ is triply ambiguous, as is $\sqrt[3]{\bar{D}}$. The three cube roots of $\bar{D}$ are the conjugates of the three cube roots of D. We want to interpret Dal Ferro's formula in such a way as to end up with specific real numbers $X = \sqrt[3]{D} + \sqrt[3]{\bar{D}}$, which are (the three) solutions of our equation $X^3 = bX + c$. The only way we might even hope to achieve this is to interpret the two cubic radicals in the sum as conjugates of one another, in which case their sum will be a real number (twice the real part of either cubic radical).

So, for each choice of cube root, $\sqrt[3]{D}$, representing the first radical in Dal Ferro's formula, let us interpret the second radical, $\sqrt[3]{\bar{D}}$, to be its conjugate. This gives us (in total) three possible interpretations of

$$X = \sqrt[3]{D} + \sqrt[3]{\bar{D}}$$

as real numbers: exactly as many values of X as our equation has solutions. *But are these three real numbers X, arrived at by this peculiar process, actually the solutions of our equation $X^3 = bX + c$?*

The answer is yes, and at this point in our story we are not that far from being able to give a full proof of this. We shall not do this (except briefly in this endnote).[2] Rather, let us sum up how (helped by a few centuries of mathematical thought) we have succeeded in interpreting Bombelli's sophistic expressions as algorithms for the computation of solutions to equations.

If the complex number $D = c/2 + \sqrt{c^2/4 - b^3/27}$ has magnitude r and phase α, then its three cube roots (call them P, Q, R) will all have the same magnitude $\sqrt[3]{r}$ and will have phases (in degrees)

$$\frac{\alpha}{3}, \frac{\alpha}{3} + 120, \text{ and } \frac{\alpha}{3} + 240.$$

Plot these three complex numbers (P, Q, R), and for each of them let X denote twice its real parts. These three real numbers X are the solutions to our equation $X^3 = bX + c$. Of course, to actually "effect" this algorithm you must be able to find the cube root of a real number ($\sqrt[3]{r}$) and you must also be able to trisect the angle α.

Nothing comes for free in this world: to solve our cubic equations, we must be able to *trisect angles.* The general angle-trisection problem (to give an algorithm for dividing any given angle into three equal angles) cannot be achieved using the classical rules of construction with straightedge and compass. But as we mentioned in section 33, the observation that there is a link between angle trisection and cubic equations had been hinted at by Bombelli, and an analogous link between "angle division" and the solution of certain polynomial equations of higher degree was perceived by Viète.

Viète managed, in 1593, to use these ideas to solve a particular polynomial equation of the 45th degree, the problem being offered as a challenge question by the Belgian mathematician Adriaen van Roomen. This challenge seems to have piqued a certain amount of national pride: "the ambassador from the Low Countries to the court of Henry IV boasted that France had no mathematician capable of solving the problem proposed by his countryman."[3] Viète succeeded by noticing that van Roomen's polynomial could be solved by cutting the circle into 45 equal parts.

Even though no national pride is now at stake, try your hand at using the algorithm discussed above to solve two cubic polynomial equations.

First, find (to an accuracy of a few decimal places) the three real numbers each having the property that its cube is equal to 3 times itself plus 1. That is, find the three solutions to the equation $X^3 = 3X + 1$.

You will first need to calulate D. Then you will need to have the approximate positions of the points P, Q, and R in the complex plane. You can compute this yourself with a protractor and a ruler, or you can make use of the diagram that follows. (If you need a further hint, this endnote explains how to find the roots.[4])

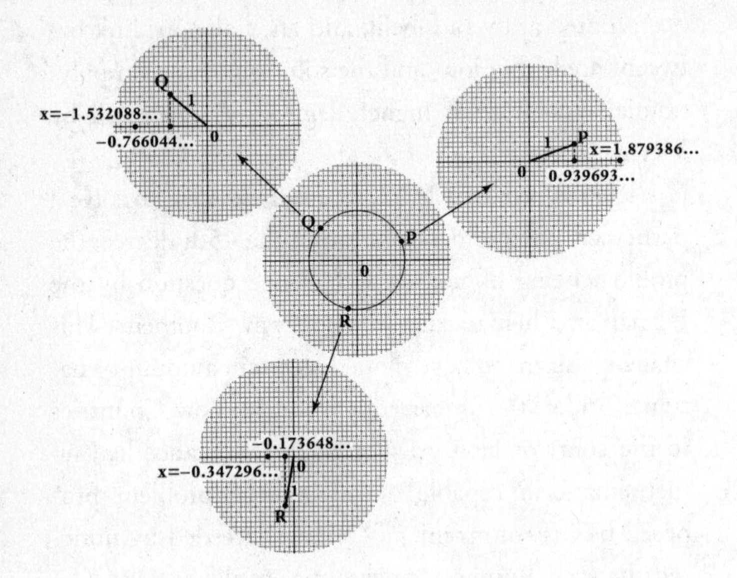

Now try your hand at an easier problem. Find the three solutions to the equation $X^3 = 15X + 4$, a problem taken up by Bombelli in this part of his manuscript:

Ancora si può procedere nella equatione di questo Capitolo in un'altro modo, come si hauesse ad agguagliare 1 a 15 p. 4 pigliasi il terzo de le Cose, che è 5 cubati fa 125, et questo si caua del quadrato della metà del numero che è 4; restarà 0 m. 121, che di questo pigliata la Radice, dirà R[0 m. 121], che aggiunta con la metà del numero, farà 2 p R[0 m 121], che pigliatone il creator Cubico, et aggiunto col suo residuo farà R[2 p R[0 m 121]] p. R[2 m. R[0 m 121]]. et tanto uale la cosa. Et benchè questo modo si possa più tosto chiamar sofistico, che altrimente come fu detto innanti nel Capitolo di Cubi, et Numero eguali a Cose, che pure nell'operatione serue senza difficoltà niuna, et assai uolte si truoua la ualuta de la Cosa per numero, come questo, che ha creatore; che'l Creatore di R[2 p R[0 m 121]] sarà 2 p R[0 m 1]; che aggiunto col suo residuo, che è 2 m R[0 m 1] che aggiunti insieme fanno 4, che è la ualuta de la cosa.

1 Eguale à 15 p 4

Soma R[0 m 121] Caua R[0 m 121]
R[2 p. R[0 m. 121]] p R[2 m R[0 m. 121]]
Creatore 2 p R[0 m. 1] p. 2 m. R[0 m. 1]
2 m. R[0 m. 1]
Somma 4: et tanto uale la Cosa

If you can't make out his handwriting, don't worry: the problem will become even easier when you are given the hint that the cube of $2 + i$ is $2 + 11i$.

64. Form and content.

We have tried in this book to replay, as a single effort of the imagination, the collective mathematical enterprise of imagining $\sqrt{-1}$. We have interleaved this replay with the reading of the phrase "the yellow of the tulip." But any parallel between these two imaginative exercises is skewed by a number of things, not the least of which is the role played by *formal structure.*

For any phrase, and, all the more, for any collection of lines in a poem, sound pattern and word arrangements are laced with formal structure: in "the yellow of the tulip," the trochee of the word *yellow* is, for example, echoed in the trochee of *tulip*, and both long-shorts are balanced on the fulcrum of the middle *l* sound. Moreover, the two words, *yellow* and *tulip*, are so similar in stress values that they are also in fine balance about the *of* in the phrase. If, in reciting that phrase, we give the slightest edge of emphasis to either of the two accented syllables, we impart a different significance to the phrase.

This type of formal architecture related to the sounds of words is important work done by any poem.[5] Words come to a poem with their sounds, demands, and habits. We can try to give a formal account of this.

As the scholar and literary critic Helen Vendler says about lyric poetry, "the true 'actors' in lyric are words, not 'dramatic persons'; and the drama of any lyric is constituted by the successive entrances of new sets of words, or new stylistic arrangements (grammatic, syntactical, phonetic) which are visibly in conflict with previous arrangements used with reference to the 'same' situation."[6]

But absolutely no formal analysis, no matter how complete, of "the yellow of the tulip" will capture the flash of color we see, the smell of that tulip, the feel of its petals, the timbre of voice in which we read the phrase to ourselves—all the imagined sensations that may tumble onto us, and surely some of them must, as we dwell with that phrase. In short, any piece of poetry is marvelously messy, if it works at all, and brings along with it a cornucopia defying formal analysis. Comprehending the formal structure of a poetic phrase may be a passport to comprehending its nature, but one then has to make the journey, tracing backward the path of the poet:

After great pain, a formal feeling comes—
The nerves sit ceremonious, like Tombs—[7]

In contrast, mathematics seems less messy, much more tied to its formal architecture, cathedral-like, crystalline. Kant's somewhat enigmatic suggestion that "mathematics has an essence, but no nature" has various possible interpretations. If, however, we take our cues from *The Critique of Judgment*,[8] we can hear in this remark Kant's mulling over the fact that mathematical truths are magnificently enclosed within their formal properties and formal consequences. And therefore, when you read Kant's dictum "Mathematics is *pure* poetry," the word *pure* (in German, *reine*), especially coming from Kant, carries an important (somewhat ironic, to my ear) twist.

Along similar lines, it is tempting to modify Oscar Wilde's assertion that "the morality of art consists in the perfect use of an imperfect medium"[9] to say that the ideal of mathematics, perhaps never achieved, is to do "perfect work in a perfect medium."

65. But . . .

I don't believe that last statement, not for a second. The real work of mathematics is hardly characterized by perfect tenses, or imperfect ones. The great glory of mathematics is its durative nature; that it is one of humankind's longest conversations; that it never finishes by answering some questions and taking a bow. Rather, mathematics views its most cherished answers only as springboards to deeper questions. I once heard Bach's harmonies described similarly: the minute Bach has established a key, this is the signal that he has already begun preparations for yet another modulation.

And with this in mind, let us return to Ashbery's lines about the inventors of writing:

> To what purpose did they cross-hatch so effectively, so that the luminous surface that was underneath is transformed into another, also luminous but so shifting and so alive with suggestiveness that it is like quicksand, to take a step there would be to fall through the fragile net of uncertainties into the bog of certainty.[10]

I take this to be a wonderful description of the experience of being-in-the-act-of-reading, where each word, as read, shimmers—perhaps even its part of speech uncertain—a surmise of the reader, who, while fighting a quicksand (which all readers of P. D. James know one should *not* do!), is at least fighting a quicksand of suggestiveness.

Ashbery's verb for the cursive scratching of writers and scribes, making marks on paper, is *cross-hatch*. Now, the most immediate visual image of what we are doing when we crosshatch is that we are drawing some gridlike figure; perhaps a *net*. A fragile net does indeed appear, later in Ashbery's sentence, and so the net image plays the curious role of being both the very source of suggestiveness of the medium and, at the same time, a somewhat ineffective protection against its teeming suggestiveness. Nets, after all, can capture and save. Nevertheless, the logic of this construction has a fizzy restlessness to it, where the medium exceeds itself, somewhat like the logic of the lamenting eulogy for Antony delivered by Cleopatra in the recitation of her dream, toward the end of Shakespeare's play (Act V, scene 2):

> . . . *His delights*
> *Were dolphin-like; they showed his back above*
> *The element they lived in. In his livery*
> *Walked crowns and crownets. Realms and islands were*
> *As plates dropped from his pocket."**

But once we read any imaginative writing, we, as readers, have already taken that dangerous step Ashbery talks of: as we imagine what we read, we are caught in, delighted by, dependent upon, that "fragile net" of suggested images, until . . . Until these images settle into something we feel we have understood; that is, until we are "bogged by certainty," and then we read on—at which point the flash of certainty is "withdrawn," "like any subtracted memories," like Keats's "joy, whose hand is ever at his lips, bidding adieu," making us ready for yet more suggestiveness.

*The inexhaustible energy of these lines comes, it seems to me, from those two "free radical" pronouns: "they" connected somehow to the simile "His delights / Were dolphin-like." Are his delights like dolphins? Or are they like the delights of dolphins? However you interpret it, or even if you don't, the meaning of these lines is vivid. No grammar can tame these lines, just as no medium can contain Antony: "His legs bestrid the ocean; his reared arm / Crested the world."

Suggestiveness and surprise are never in short supply in mathematics. Success in establishing, and then comprehending, Dal Ferro's formula for the solution of the general cubic equation, and success in discovering a similar formula for the solution of the general fourth-degree equation—the solution being given by the extraction of fourth roots and cube roots—suggest that one could seek an overarching formula to find the solutions of the general polynomial equation of any degree, the solutions being given by the extraction of appropriate roots. Here is the surprise: there is no such overarching formula, and, moreover, one can prove its nonexistence (even for the next case: polynomial equations of the fifth degree). Now, to establish a formula is one thing, but to establish the *nonexistence* of a formula is a chore of quite a different magnitude. This task, started at the beginning of the nineteenth century by the Italian mathematician Paolo Ruffini and continued by the Norwegian mathematician N. H. Abel, invites the mathematical imagination to an utterly new type of play and offers a cornucopia of new suggestiveness.

But that would be another book.

The Poems of Our Climate

I

Clear water in a brilliant bowl,
Pink and white carnations. The light
In the room more like a snowy air,
Reflecting snow. A newly-fallen snow
At the end of winter when afternoons return.
Pink and white carnations—one desires
So much more than that. The day itself
Is simplified: a bowl of white,
Cold, a cold porcelain, low and round,
With nothing more than the carnations there.

II

Say even that this complete simplicity
Stripped one of all one's torments, concealed
The evilly compounded, vital I
And made it fresh in a world of white,
A world of clear water, brilliant edged,
Still one would want more, one would need more,
More than a world of white and snowy scents.

III

There would still remain the never-resting mind,
So that one would want to escape, come back
To what had been so long composed.
The imperfect is our paradise.
Note that, in this bitterness, delight,
Since the imperfect is so hot in us,
Lies in flawed words and stubborn sounds.

Wallace Stevens[11]

APPENDIX: THE QUADRATIC FORMULA

The "method" for solving quadratic equations,

$$X^2 + bX + c = 0,$$

is called *completing the square*, and it will lead us to a specific formula giving the two solutions (in general there are two). For fun, let me rehearse one way to "derive" the formula.

The opening move, as is often the case in mathematical arguments, is to (provisionally) assume the existence of a solution X to the problem at hand, and see what properties X must have if it is indeed a solution. In the best of circumstances, this leads to a complete description of the X that is sought for. The end move is to show that the X found by this procedure is truly a solution to the problem. Such opening moves and end moves, tagged in some of the classical literature as *analysis* and *synthesis*, respectively, were sometimes thought of as mirror images of one another, the *synthesis* legitimizing the *analysis*. Let us perform, then, an *analysis* (in the above sense) of our problem.

Replace X (the unknown we want to find) in the equation

$$X^2 + bX + c = 0$$

by the expression $Y - b/2$ (if we can find Y we can immediately find X). You will soon see why the method is called "completing the square."

The equation obtained after this substitution,

$$\left(Y - \frac{b}{2}\right)^2 + b\left(Y - \frac{b}{2}\right) + c = 0,$$

simplifies to

$$\left(Y^2 - bY + \frac{b^2}{4}\right) + \left(bY - \frac{b^2}{2}\right) + c = 0,$$

or

$$Y^2 = \frac{b^2}{4} - c = \frac{b^2 - 4c}{4}.$$

That is, our equation, which was initially a somewhat complicated equation in the unknown X, now involves the new unknown Y only in its "square term" (Y^2): we have completed the square. So,

$$Y = \frac{\sqrt{b^2 - 4c}}{2} \quad \text{or} \quad Y = -\frac{\sqrt{b^2 - 4c}}{2},$$

and therefore, since X is $Y - b/2$, we get the *quadratic formula*:

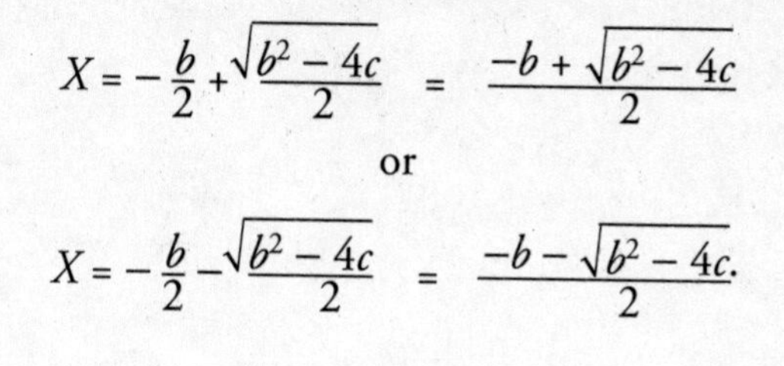

$$X = -\frac{b}{2} + \frac{\sqrt{b^2 - 4c}}{2} = \frac{-b + \sqrt{b^2 - 4c}}{2}$$

or

$$X = -\frac{b}{2} - \frac{\sqrt{b^2 - 4c}}{2} = \frac{-b - \sqrt{b^2 - 4c}}{2}.$$

NOTES

Preface

1. See, e.g., Karen Wynn, "Addition and Subtraction by Human Infants," *Nature* (1992) 358: 749–50.
2. Elaine Scarry, in her wonderful essay "Imagining Flowers: Perceptual Mimesis (Particularly Delphinium)" (*Representations* [1997] 57: 90–115), had suggested this was a difficult or impossible thing to do—which we took as a challenge. A later version of Scarry's essay appears as the chapter "Imagining Flowers" in her *Dreaming by the Book* (Farrar, Straus and Giroux, 2000), pp. 40–74.

1. The Imagination and Square Roots

1. I thank David Gewanter for pointing this out to me.
2. Scarry, *Dreaming by the Book*, p. 42.
3. This phrase comes from John Ashbery's prose poem "Whatever It Is, Wherever You Are," in his *A Wave: Poems by John Ashbery* (Noonday Press/Farrar, Straus and Giroux, 1985), pp. 63–65.
4. From Rilke's letter to his Polish translator, Witold von Hulewicz, a portion of which is quoted in Rainer Maria Rilke, *Duino Elegies*, trans. B. Leishman and S. Spender, 4th ed., rev. (Hogarth Press, 1963), p. 157. Here is a bit more of Leishman and Spender's translation: "Our task is to stamp this provisional, perishing earth into ourselves so deeply, so painfully and passionately, that its being may rise again, 'invisibly,' in us. We are the bees of the invisible."

5. I am recalling this, perhaps imperfectly, from an interview that I heard on NPR.
6. From Rilke's letter to his wife, in *Letters of Rainer Maria Rilke*, trans. Jane Bannard Greene and M. D. Herter Norton (Norton, 1945), p. 286.
7. A recent English translation is in Plato, *Complete Works*, ed. J. M. Cooper (Hackett, 1997), pp. 870–97.
8. Students of the calculus will have seen the interpretation of integrals as "signed areas" of certain regions of the plane. But this notion doesn't respond to our Zen koan: the *signed area* of such regions is simply the difference between the area of the portion of the region above the horizontal axis and the area of that portion below; it is positive or negative, depending on whether the major part of the region lies above or below the horizontal axis.
9. If you can multiply small numbers and think about the operation as you do it, if you can substitute values in simple algebraic expressions, and if you can gaze on algebraic expressions with equanimity, you shouldn't have very much trouble with the mathematics in this book. Here are samples of these three activities.

   (1) *Multiplication*. Let us multiply 23 times 45 and watch what we are doing.

$$\begin{array}{r} 45 \\ \times\, 23 \\ \hline 135 \\ +\, 90 \\ \hline 1035 \end{array}$$

   Think about the significance of the 135 and 90: implicit in this piece of reckoning is that we have parsed 23 as "3 ones and 2 tens" ($23 = 3 + 20$) and have broken up the task of multiplying 23 by 45 into the two partial tasks of multiplying 3 by 45 (giving the 135) and 20 by 45 (giving, as written above, 90 tens). We then add up, banking on this equation:

$$(3 + 20) \times 45 = 3 \times 45 + 20 \times 45.$$

We will see more of this kind of equation.

(2) *Substitution.* We will be substituting "values" for "unknowns" in algebraic expressions. For example, substituting $X = 1$, $X = 2$, $X = 3$, and $X = 4$ for the unknown X in the expression $X^2 - 2X + 1$, we get

$$1^2 - 2 \times 1 + 1 = 0$$
$$2^2 - 2 \times 2 + 1 = 1$$
$$3^2 - 2 \times 3 + 1 = 4$$
$$4^2 - 2 \times 4 + 1 = 9.$$

Do you see the pattern? The value of the expression $X^2 - 2X + 1$ is always the square of 1 less than the value you substituted for X. Or, $X^2 - 2X + 1 = (X - 1)^2$.

(3) *Equanimity.* Take a look right now at the most complicated algebraic expression that will appear in this book (Dal Ferro's formula, on p. 117), and don't turn away. If you are puzzled by this curious expression, you are in the company of sixteenth-century algebraists. We will be discussing its importance, parsing it, and eventually turning it to our use in the pages that follow.

10. Dennis Sullivan, "The Density at Infinity of a Discrete Group of Hyperbolic Motions," *Publications Mathématiques de l'IHES* (1979) 50: 419–50.
11. Paul Scott, *The Jewel in the Crown* (William Morrow, 1966); this is the first book of Scott's *The Raj Quartet.*
12. Eva Brann, *The World of the Imagination—Sum and Substance* (Rowman & Littlefield, 1991).
13. Quintilian, *Institutio Oratoria*, vol. 6, p. 2; trans. in Brann, *World of Imagination*, p. 21.
14. For a discussion of Jeremy Bentham's *Essay on Logic*, see Brann, *World of Imagination*, p. 23.
15. Wordsworth, as reported by his friend Henry Crabb Robinson; for a discussion of this, see Mary Warnock, *Imagination* (Univ. of California Press, 1976), p. 114.
16. From Brann, *World of Imagination*, p. 23 (but the sentiment behind this quotation is not Brann's!).
17. This is in chapter 13 of the first volume of the *Biographia Lit-*

eraria, in Coleridge, *Selected Poetry and Prose,* ed. S. Potter (Nonesuch, 1971), p. 246.

18. Alexander Stille, "The Betrayal of History," *New York Review of Books,* June 1998.
19. Ashbery, "Whatever It Is."
20. Elaine Scarry, *Dreaming by the Book,* p. 42.
21. This is the answer to the following problem: Given a triangle whose area is 3 and whose sides are of length $X - 1$, X, and $X + 1$, what is X? It appears in G. Cardano, *Artis Magnae Sive de Regulis Algebraicis,* the first edition of which was published in 1545. All translations of this text quoted here are from T. Richard Witmer, ed. and trans., *Cardano: The Great Art* (MIT Press, 1968), hereafter referred to as *Ars Magna.* I will also be quoting from Oystein Ore's foreword to this volume. The problem above is posed and solved on page 193 of *Ars Magna.*
22. The problem quoted in the text is Problem VII in *Ars Magna,* pp. 44–45. Cardano works it out by reducing the problem to the solutions of a quadratic equation and getting the answer that 100 aurei will pay for 7 horsemen or 25 foot soldiers.
23. H. T. Colebrooke, *Algebra with Arithmetic and Mensuration from the Sanscrit of Brahmegupta and Bháscara* (John Murray, 1817), p. 211. I am thankful to Manjul Bhargava for consulting Bháskara's original text and for assuring me that this word-problem has this unambiguous interpretation: If n is the number of bees, then n is equal to $\sqrt{n/2} + 8n/9 + 2$. This leads to the equation $(n - 72)(2n - 9) = 0$, giving only one possible answer for the number of bees in the hive. There are 72 of them.
24. See, in particular, Rafael Bombelli, *L'Algebra, prima edizione integrale,* ed. E. Bortolotti and U. Forti (Feltrinelli, 1969).
25. See the discussion of Chuquet's manuscript in F. Cajori, *A History of Mathematical Notations* (Open Court Publ., 1929), vol. 2, p. 126, para. 495. In chapter 2 we will be reviewing the *quadratic formula,* from which you can see how Chuquet might have been led to an "impossible" answer to his problem.
26. Squaring $3/2 + \sqrt{-7/4}$ (i.e., multiplying it by itself "by the ordinary laws of arithmetic") gives a sum of four terms,

$$\left(\frac{3}{2}+\sqrt{\frac{-7}{4}}\right)\times\left(\frac{3}{2}+\sqrt{\frac{-7}{4}}\right)=\frac{9}{4}+\frac{3}{2}\times\sqrt{\frac{-7}{4}}+\frac{3}{2}\times\sqrt{\frac{-7}{4}}+\frac{-7}{4},$$

which simplifies to $1/2 + 3\sqrt{-7/4}$. The "ordinary law of arithmetic" that we have used to perform this calculation,

$$(a + b)(c + d) = ac + ad + bc + bd,$$

is a variant of something called the *distributive law*, which will be one object of our attention in chapter 2.

27. *The Riverside Chaucer*, ed. Larry D. Benson (Houghton Mifflin, 1987), p. 135.
28. The statement of the Poincaré conjecture is usually given in the language of *topology*, that branch of mathematics sometimes referred to in popular expositions as "rubber geometry": two spaces are considered "topologically equivalent" if, thinking of both of them as made of rubber, you can stretch or compress or twist one of them to look like the other—but no tearing is allowed (unless you then mend the tear perfectly). The Poincaré conjecture is the statement that the following is a topological characterization of Euclidean three-dimensional space: Any "space" X (let us request that it be a metric space) that has the following three properties is topologically equivalent to Euclidean three-dimensional space. (1) X is *locally Euclidean* in the sense that each of its points has a small surround that is topologically equivalent to a small surround of a point in Euclidean three-dimensional space. (2) X is *Euclidean at a far distance* in the sense that there is a region in X that incorporates all points sufficiently far from one of its points, and this region is topologically equivalent to some region that has the same description (it incorporates all points sufficiently far from one point) in Euclidean three-dimensional space. (3) The entire space X can be continuously shrunk to a single one of its points, the shrinking process never leaving X (a kind of reverse Big Bang).

2. Square Roots and the Imagination

1. The proof we are about to give is not the only one, and probably not the oldest. For a discussion of the history of this problem, see David Fowler, *The Mathematics of Plato's Academy* (Clarendon Press, 1987), pp. 1–30.
2. Ibid., p. 73.
3. The expression $X^2 - 3X + 4$ is an example of a *quadratic polynomial* (in the unknown X), and, to take another random example, so is the expression $(5/2)X^2 + (0.99)X - 1$. Quadratic polynomials are the sum of three terms, the *square term* (X^2 in our first example, $(5/2)X^2$ in our second), the *linear term* ($-3X$ in our first example, $(0.99)X$ in our second), and the *constant term* (4 in our first, -1 in our second). The general quadratic polynomial can be thought of as an expression $aX^2 + bX + c$; a choice of specific numerical values for a, b, and c will give a specific polynomial. These numerical values a, b, and c are called the *coefficients* of the polynomial. For a polynomial to be genuinely quadratic (rather than linear), its *leading coefficient*, a, must not be zero. The adjective *quadratic* refers to the fact that the highest power of the unknown that occurs in the expression is the square term (the noun "square" being *quadratum* in Latin). We shall also be dealing later with *cubic polynomials*, a random example of which is $X^3 + 2X^2 + 3X + 4$ (*cubic*, because the highest power of the unknown that appears in this expression is 3). A *zero*—or, synonymously, a *root*—of a polynomial is a number ν that when substituted for the unknown X in the polynomial expression gives the value 0. So, a root ν of the polynomial $X^3 + 2X^2 + 3X + 4$ is a number such that $\nu^3 + 2\nu^2 + 3\nu + 4 = 0$.
4. Virginia Woolf, *To the Lighthouse* (Harvest, 1955), pp. 53–54; *On Beyond Zebra* is the children's book by Dr. Seuss (Random House, 1980).
5. Colebrooke, *Algebra*, p. 135.
6. Isaac Newton, *Universal Arithmetick*, trans. Ralphson, 2d ed. (Senex-Innys, 1728), p. 197. Newton wrote his treatise in Latin.
7. This is from Diaconis's introduction to a new Italian transla-

tion of Cardano's work: G. Cardano, *Liber de Ludo Aleae*, trans. G. Gammbacorta (Univ. of Pavia Press, 2002). See also Oystein Ore's *Cardano, The Gambling Scholar* (Dover, 1965) and the more recent biography by Anthony Grafton, *Cardano's Cosmos: The World and Works of a Renaissance Astrologer* (Harvard Univ. Press, 1999).

8. *Ars Magna*, pp. 1, 8.
9. For a discussion of this, see Fowler, *Mathematics of Plato's Academy*, pp. 3–8.
10. Moses Maimonides, *The Guide of the Perplexed*, trans. Shlomo Pines (Univ. of Chicago Press, 1963), vol. 1, p. 6. Maimonides is referring to "Tractate Hagigah" (chap. 2 of Mishnah 1): "The forbidden degrees may not be expounded before 3 persons nor the Story of Creation before 2 nor [the chapter of] the Chariot before 1 alone, unless he is a Sage that understands of his own knowledge" (*The Mishnah*, trans. Herbert Dandy [Oxford Univ. Press 1933], pp. 212–13). I am thankful to Avner Ash for this reference.
11. See footnote in *Ars Magna*, p. 220.
12. *Ars Magna*, p. 219.
13. Ibid., p. 219, footnote. See also V. Sanford's translation of the passage in D. E. Smith, *A Source Book in Mathematics* (Dover, 1984), vol. 1, p. 202.
14. Bombelli, *L'Algebra*.
15. Bombelli abbreviated *più di meno* to p.d.m., which allowed him to refer to, say, −2 as p.d.m.2. For further discussion, see Federica La Nave and Barry Mazur, "Reading Bombelli," *Mathematical Intelligencer*, Jan. 2002, pp. 12–20.

3. Looking at Numbers

1. This image is from Ludwig Wittgenstein, *Philosophical Investigations*, trans. G.E.M. Anscombe, 3d ed. (Macmillan, 1968), para. 106:

> Here it is difficult as it were to keep our heads up, — to see that we must stick to the subjects of our everyday thinking, and not go astray and imagine that we have to

describe extreme subtleties, which in turn we are after all quite unable to describe with the means at our disposal. We feel as if we had to repair a torn spider's web with our fingers.

2. Brann, *World of the Imagination*, pp. 3, 32.
3. For a discussion of the Stoic view of the imagination, see Brann, pp. 46–48. For a spirited discussion of the the Stoic notion of "impression" see Sextus Empiricus, Vol. II, *Against the Logicians* (vol. 2 in Loeb Classical Library), trans. R. G. Bury (Harvard Univ. Press, 1935), pp. 122–41 (ll. 227–62) and 196–207 (ll. 372–87). For Ibn al-'Arabī, see William Chittick, *Ibn al-'Arabī's Metaphysics of Imagination: The Sufi Path of Knowledge* (State Univ. of New York Press, 1989), p. 16. One can understand the delay in full publication once one learns from Osmon Yahia's *Histoire et classification de l'oeuvre d'Ibn 'Arabī* (Institut Français de Damas, 1964) that the critical edition of the *Futūhāt* is estimated to run to 17,000 pages.
4. W. B. Yeats, "Adam's Curse," in *The Collected Poems of W. B. Yeats* (Scribner, 1996), p. 80.
5. John Livingston Lowes, *The Road to Xanadu* (Houghton Mifflin, 1930).
6. S. T. Coleridge, *Selected Poetry and Prose* (Nonesuch Press, 1971), p. 93.
7. Paula Panich, personal communication.
8. Rainer Maria Rilke, *Letters to a Young Poet*, trans. Stephen Mitchell (Random House, 1987), pp. 23–25.
9. See, e.g., A. L. Cauchy's paper in *Comptes Rendus Acad. Sci.* 11 (1847): 1120.
10. Augustus De Morgan, *Trigonometry and Double Algebra* (London, 1849), p. 41. See the discussion of this in Cajori, *History of Mathematical Notations*, pp. 130–31, para. 501.
11. Stendhal, *The Life of Henry Brulard*, trans. Jean Stewart and B.C.J.G. Knight (Merlin Press, 1958), chap. 33. The quotations in the text are from pp. 257, 258, and 260 of this volume. See also the translation by John Sturrock, *Stendhal: The Life of Henry Brulard* (Penguin, 1995), pp. 355–58.
12. This is from Anna Pavord, *The Tulip, The Story of a Flower That*

Made Men Mad (Bloomsbury, 1999), p. 27; the color plates in this volume are extraordinary. Another recent work on the tulip and its history is Mike Dash, *Tulipomania* (Random House, 1999).

13. Dash, *Tulipomania*, p. 7.
14. For a discussion of this see Pavord, *Tulip*, chap. 1 (esp. p. 35); and Dash, *Tulipomania*, p. 19.
15. Dash, *Tulipomania*, p. 20.
16. Pavord, *Tulip*, p. 43.
17. Dash, *Tulipomania*, p. 10; Pavord, *Tulip*, p. 33.
18. This is from the poem "The Old Vicarage, Grantchester."
19. There is one discrepancy: Rimbaud associates the letter *E* with white, whereas in the primer, *E* is yellow. For a discussion of this see Enid Starkie, *Arthur Rimbaud* (Greenwood Press, 1978), p. 165.
20. For an impressive array of such gauged instruments available by 1700, see fig. 2.19 of P. H. Sydenham, *Measuring Instruments: Tools of Knowledge and Control* (Peter Peregrinus, 1979).
21. Colebrooke, p. 132.
22. Barbara Tversky, "Cognitive Origins of Graphic Productions," chap. 4 in *Understanding Images,* ed. F. T. Marchese (Springer-Verlag, 1995), pp. 29–53. See also Tversky's survey article "Spatial Schemas in Depictions," in *Spatial Schemas and Abstract Thought*, ed. M. Gattis (MIT Press, in press).
23. From Mark Johnson and George Lakoff, *Metaphors We Live By* (Univ. of Chicago Press, 1980). I am also thankful to George Lakoff and to Barbara Tversky for their help and their correspondence about these matters.
24. John Napier (1550–1617) is credited with the discovery of logarithms. For a brief discussion of his "kinematic conception of logarithms," see Ivor Grattan-Guinness, *The Norton History of the Mathematical Sciences* (Norton, 1997), p. 218.
25. This was related to me by Robert Kaplan, who with his wife, Ellen Kaplan, runs the Math Circle, an extracurricular math enrichment program in the Boston area.
26. For a recent translation, see E. Brann, P. Kalkavage, and E. Salem, trans., *Plato's Sophist* (Focus Philosophical Library, 1996).

27. For this, see Definition 5 of Book 5 of *Euclid's Elements.* See also T. L. Heath's extensive note on this, which discusses, among other things, the possibility that Eudoxus is the author of this definition, in T. L. Heath, trans. and commentary, *Euclid's Elements* (Dover, 1956), vol. 2, pp. 120–26.
28. For example, truncating the infinite continued fraction (see sect. 6)

$$1 + \cfrac{1}{2 + \cfrac{1}{2 + \cfrac{1}{2 + \dots}}}$$

at successive finite points gives a series of fractions that close in on 2 alternately from above and below:

$$1 < \sqrt{2}$$

$$1 + \frac{1}{2} = \frac{3}{2} > \sqrt{2}$$

$$1 + \cfrac{1}{2 + \cfrac{1}{2}} = \frac{7}{5} < \sqrt{2}$$

See Fowler, *Mathematics of Plato's Academy* (Clarendon Press, 1987), for some interesting historical speculations about the antiquity of this mode of definition of (real) number.
29. For an excellent, not too technical account of this, see Tobias Dantzig, "The Art of Becoming," chap. 8 in *Number: The Language of Science* (Free Press, 1954), esp. pp. 139–63.

4. Permission and Laws

1. This is what I remember of an interview with Gabriel García Márquez that I heard on the radio many years ago. I haven't been able to track down that interview to verify my memory of it, but García Márquez has commented about *The Metamorphosis* in many places. He is reported to have said that Kafka wrote (specifically in the first sentence of *The Metamorphosis*) "the way grandmother (*abuela*) used to talk"; and "Damn, I did not know that such a thing could be done!"; and that if this is allowed, "then writing interests me." Contrast this with Gar-

cía Márquez saying, as well, "One cannot write whatever one feels like, there are laws which must be respected." For a fuller report of all this, see Plinio Mendoza, *El Olor de la Guayaba: Conversaciones con Gabriel García Márquez* (Diana, 1982). I am thankful to Sorana Froda for providing this reference. The translation of the first sentence of Kafka's *The Metamorphosis* is by Willa and Edwin Muir, as cited in Ann Charters, *The Story and Its Writer* (St. Martin's Press, 1995), p. 733.

2. *Ars Magna,* p. 9.
3. Diophantus is generally considered to have lived in the third century A.D., but see the discussion on pp. 171–72 of O. Neugebauer, *The Exact Sciences in Antiquity* (Princeton Univ. Press, 1952).
4. *Ars Magna,* p. 9.
5. Unfortunately, in my case, it must be the (English) translation of these texts to which I turn.
6. Here, as Manjul Bhargava assures me, Bháskara means that if n is the number of geese, n is equal to $10\sqrt{n} + n/8 + 6$, which leads to the equation $(n - 144)\ (49n - 16) = 0$, telling us that there must have been 144 geese in the flock.
7. Viète's *In Artem Analyticem Isagoge* (Introduction to the Analytic Art) is a highly detailed programmatic sketch of a longer work (*Algebra Nova*) that never appeared in Viète's lifetime. For a discussion of Viète's work, see J. Klein, *Greek Mathematical Thought and the Origin of Algebra*, trans. E. Brann (MIT Press, 1968), pp. 151–53. For the *Isagoge*, see *Francisci Vietae* [François Viète], *Opera Mathematica* (F. van Schooten, 1646), pp. 1–12. For an English translation of the *Isagoge*, see J. Winfree Smith, Appendix, in Klein, *Greek Mathematical Thought*, pp. 315–53. See also François Viète, *The Analytic Art*, trans. T. Richard Witmer (Kent State Univ. Press, 1983).
8. *The Analytic Art*, p. 26.
9. Ibid.
10. A modern algebraist would not make such a claim. Since Viète doesn't allow his unknowns to take zero as a value, he is okay.
11. The capital letters are Viète's: "fastuosum problema problematum ars Analytice . . . iure sibi adrogat, Quod est, NULLUM NON PROBLEMA SOLVERA."

12. I found Fichte's statement in David Lachterman, *The Ethics of Geometry* (Routledge, 1989), p. 17; the reference is Fichte, "Über das Verhältnis der Logik zur Philosophie oder transcendentale Logik," in *Nachgelassene Schriften,* Bd. 9, pp. 42–43.
13. The term *distributive* was introduced by J.-F. Servois in two articles published in 1814 in the *Annales des Mathématiques.* It would seem from these articles that his motivation for formulating the term is to be able to discuss (what later mathematicians would call) linear (differential) operators; e.g.,

$$L(f+g) = L(f) + L(g).$$

14. This is a paraphrase of Euclid's First Postulate as given, e.g., in Proclus, *Commentaries on Euclid*, the relevant discussion (in Thomas Taylor's translation) beginning with the sentence "Let it be granted that a straight line may be drawn from any one point to any other point." See Thomas Taylor, *The Philosophical and Mathematical Commentary of Proclus on the First Book of Euclid's Elements* (T. Paine & Sons, B. White & Sons, J. Robson, T. Cadell, Leight & Co., G. Nichol, R. Foulder, T. & J. Eggerton, 1792), vol. 2, pp. 6–8.
15. Ernst Mach, *The Science of Mechanics*, trans. T. J. McCormack (Dover, 1984).
16. The poem "Elegy," originally published in W. S. Merwin's 1970 volume *The Carrier of Ladders*, is reprinted in W. S. Merwin, *The Second Four Books of Poems* (Copper Canyon Press, 1993) p. 226.

5. Economy of Expression

1. René Descartes, "Discours de la Méthode," 2d part, in his *Oeuvres philosophiques* (1618–37), Tome 1 (Garnier Frères, 1963), p. 589.
2. In this discussion I am using and quoting from Marshall Clagett, *Nicole Oresme and the Medieval Geometry of Qualities and Motions* (Univ. of Wisconsin Press, 1968), pp. 165–68. On the difficulty of using the word *precursor* in this case (and oth-

ers), see Clagett's commentary in the introduction to his volume.

3. See the discussion about this in S. J. Gould, *The Mismeasure of Man* (Norton, 1981); see also P. B. Medawar, "Unnatural Science," *New York Review of Books*, Feb. 1977, pp. 13–18.
4. Jorge Luis Borges, "The Metaphor," in his *This Craft of Verse* (Harvard Univ. Press, 2000), pp. 21–42.
5. This is Coleman Barks's translation, from his *The Essential Rumi* (HarperCollins, 1996), p. 272.
6. From Robert Herrick's "A Meditation for His Mistress."
7. Chase Twichell, "Tulip," in her *The Snow Watcher* (Ontario Review Press, 1998).

6. Justifying Laws

1. A friend of mine uses this exercise effectively to raise the issue of commutativity when teaching very young children.
2. Herman Melville, *Moby Dick* (Penguin, 1985), p. 144.
3. "Creeping strategy" is my neologism for what is usually called "definition by (mathematical) induction."
4. For a discussion, from a cognitive science perspective, of the nature of arithmetic laws, see George Lakoff and Rafael Núñez, "Where Do Laws of Arithmetic Come From?" (chap. 4 in their *Where Mathematics Comes From* [Basic Books, 2000]).

7. Bombelli's Puzzle

1. *Ars Magna,* chap. 11, p. 96.
2. Since the polynomial $X^3 - 6X - 40$ has $X = 4$ as root, it has $X - 4$ as factor, so $X^3 - 6X - 40 = (X - 4)(X^2 + 4X + 10)$ and the other two roots are the roots of the quadratic factor $X^2 + 4X + 10$.
3. In sect. 33, we too will have the formula to deal with this problem. The first two dozen digits of the unique solution are $X = 1.324717957244746025960908\ldots$
4. See an account of this in Ore, foreword to *Ars Magna,* pp. vii–xiii.

5. Ibid., pp. xx–xxi. One might wonder how Fiore could have "suffered a humiliating defeat" at the hands of Tartaglia and yet have known the method rediscovered and used by Tartaglia to defeat him. Later we will consider various forms of cubic equations and how the method in question treats them. For the more modern reader, a major indicator of qualitative distinction between various cubic polynomials is how many real roots they have. The method "available" to Fiore and Tartaglia could effectively treat cubic polynomials having a single real root. But even there, things were very tricky for those contestants. Since negative numbers were somewhat avoided, it was not yet common practice to bring all terms of the cubic equation to one side of the equality sign, putting a zero on the other. And since missing terms had not yet been construed as terms with "zero as coefficient," cubic equations took thirteen seemingly different forms, each with its own (seemingly) specific method of solution. There being no single "method" (or, rather, no comprehension that the method can simply be phrased so as to apply to all thirteen forms of cubic equations), I imagine that even Fiore might mess up in applying the appropriate variant of the method to a challenge equation presented by Tartaglia.
6. *Ars Magna*, p. x.
7. Ibid, p. xi.
8. Cardano begins chapter 1 of his *Ars Magna* with this tribute to al-Khwārizmī: "This art originated with Mahomet the son of Moses the Arab" (p. 7). In contrast, in the preface to *L'Algebra* (p. 8), Bombelli refers to al-Khwārizmī as "the author of a minor work, not of great value." In this dismissive judgment, Bombelli may have been overly influenced by his new discovery, with Antonio Maria Pazzi, of seven books of Diophantus. Bombelli and Pazzi translated five of these seven books, and in the course of this labor, Bombelli says that he "came to know that this discipline [algebra] had been known to the Indians before the Arabs" (ibid., p. 8).
9. In particular, see Roshdi Rashed, "L'algèbre," in his *Histoire des sciences arabes* (Le Seuil, 1997), vol. 2.
10. The dedicatee of my book tells me that if I do three things in rapid succession—say *al-jabr wa al-muqābalah* out loud fast,

remember that Lewis Carroll was a mathematician, and note that in its first appearance in *Through the Looking Glass*, the word *Jabberwocky* is, like Arabic, written from right to left—I might believe as she does that the name *Jabberwocky* was derived from the title of that ninth-century work.

11. This is from the introduction to Book Three of Bombelli's *L'Algebra*, p. 317. The translation of this passage given in the text is Michelle Sharon Jaffe's (e-mail communication).
12. Here, I am expressing these formulas in modern terms. For one thing, Bombelli was loath to work with negative numbers and would arrange it so that all negative quantities were transposed to the opposite side of the equation; this operation he referred to as *transmutatione*.
13. I chose the word *indicator* (a neologism in this context) to emphasize that its sign "indicates" how many real solutions the cubic equation has: if the indicator is positive, the equation only has one real solution; if negative, three. The indicator is $-1/108$ times the classical *discriminant* of the equation.
14. The number -1 is a cube root of -1, so interpreting both cube roots of -1 in Dal Ferro's formula as -1, we find that the enigmatic formula predicts $X = -1 + -1 = -2$ as a solution of our equation. This gives one, at least, of the two solutions. The solutions are the roots of the polynomial

$$X^3 - 3X + 2 = (X - 1)^2 (X + 2).$$

15. The quotation is from Bombelli, *L'Algebra*, p. 133 (see chap. 1, n. 19); for the translation and discussion, see La Nave and Mazur, "Reading Bombelli," pp. 14–17 (see chap. 2, n. 10).
16. Ibid., pp. 639–42.
17. *The Analytic Art*, p. 445.
18. For an analysis of events in the history of algebra, after Bombelli, that mark such a shift to a more modern temperament, see Klein, *Greek Mathematical Thought.*
19. Albert Girard, *Invention nouvelle en l'algèbre: tant pour la solution des équations, que pour recognoistre le nombre des solutions q'elles reçoivent avec plusieurs choses que sont nécessaires à la perfection de ceste divine science* (Guillaume Iansson Blaeuw, 1629).

20. Cardano uses the words *res*, *positio*, or *quantitas*. The word *cos*, it seems, is the origin of the phrase "Rule of the coss," which was an early label for algebra.
21. Bombelli, *L'Algebra*, p. 155; translation by Michelle Sharon Jaffe in her *The Story of O* (Harvard Univ. Press, 1999), p. 47.
22. Colebrooke, *Algebra*, p. 139, para. 17 (see chap. 1, n. 18). See also the footnote on that page of Colebrooke.
23. Newton, *Universal Arithmetick* (see chap. 2, n. 3).
24. G. W. Leibniz, "Of Universal Synthesis and Analysis; or, Of the Art of Discovery and of Judgement," trans. M. Morris and G.H.R. Parkinson, in *Leibniz Philosophical Writings*, ed. G.H.R. Parkinson (Dent & Sons, 1973), pp. 10–17.

8. Stretching the Image

1. This quotation is from John Updike's excellent essay "Kafka and the Metamorphosis," in *The Story and Its Writer*, ed. Ann Charters, 4th ed. (St. Martin's Press, 1995), pp. 1502–5, in which Updike does indeed *imagine* the insect without *visualizing* it. See also the video *Nabokov on Kafka: Understanding "The Metamorphosis"* (Monterey Home Video, 1991), a reenactment of one of Nabokov's lectures on *The Metamorphosis,* with Christopher Plummer playing Nabokov.
2. Vladimir Nabokov, *Lectures on Literature* (Harcourt Brace, 1980), p. 258.
3. Ibid., p. 258.
4. The phrase is from David Magarshack's translation, reprinted in *The Anchor Book of Stories* (Doubleday & Co., 1958), p. 59. I am thankful to Don Fanger for informing me that there is indeed such a movie (which I haven't yet managed to see).
5. *Dreaming by the Book*, p. 43.
6. R. Descartes, *Regulae ad Directionem Ingenii* [Rules for the Direction of the Natural Intelligence], trans. G. Heffernan (Rodopi, 1998).
7. Winston Churchill, *The Gathering Storm* (Houghton Mifflin, 1948), p. 343.
8. The three numbers whose cube is −1 (i.e., the three *cube roots* of −1) are −1, $(1 + \sqrt{-3})/2$, and $(1 - \sqrt{-3})/2$.

9. Yes, sixfold. The nine possible choices give rise to six different complex numbers.
10. Interpreting each $\sqrt[3]{-1}$ as -1 gives $X = -2$, and interpreting one $\sqrt[3]{-1}$ as $(1 + \sqrt{-3})/2$ and the other as $(1 - \sqrt{-3})/2$ gives the solution $X = 1$. All other choices lead to nonreal numbers and are not solutions to $X^3 = 3X - 2$.
11. To get the flavor of Bombelli's volatile style, see his foreword to *L'Algebra* or the translation of it in La Nave and Mazur, "Reading Bombelli."
12. As we shall see in our discussion of De Moivre's work in chapter 11, these concoctions of roots of sums of roots may be perfectly interpreted as "numbers" of the type known to Cardano.

9. Putting Geometry into Numbers

1. A. B. Lord, *The Singer of Tales* (Harvard Univ. Press, 1960). Cf. G. Nagy, *Homeric Questions* (Univ. of Texas Press, 1996), p. 70.
2. The story in Ferdowsī's *Shāh-nāmeh* I, 21.126–36, is recounted in Nagy, *Homeric Questions*, p. 70.
3. "Wherever It Is, Whatever You Are", in *A Wave*, p. 63.
4. Adrienne Rich, "(Dedications)," in her *An Atlas of a Difficult World: Poems 1988–1991* (Norton, 1991).
5. See Cajori's discussion of this in his *History of Mathematical Notations*, p. 126, para. 495.
6. The primer in question is Robert Recorde, *The Grou[n]d of artes teachying the worke and practise of Arithmetike, moch necessary for all states of men. After a more easyer & exacter sorte, then any lyke hath hytherto ben set forth: with dyurse newe additions* (first printing, 1542). Michele Sharon Jaffe quotes this passage in her *The Story of O* (Harvard Univ. Press, 1999), p. 39. The point is that, as this piece of dialogue shows, the change from Roman to Arabic numerals effected a kind of defamiliarization that made the characters themselves akin to unknowns, things to be "evaluated" by comprehending the implicit and explicit conventions of their use.
7. See Stephen Booth, *Shakespeare's Sonnets* (Yale Univ. Press, 1977), pp. 447–52.

8. For a fine discussion, from a cognitive science perspective, of the conceptualization of *i* and the geometrization of complex numbers, see Lakoff and Núñez, "Case Study 3: What Is *i*?," in their *Where Mathematics Comes From*, pp. 420–32.
9. Thomas Lux, "The Voice You Hear When You Read Silently," in his *New and Selected Poems 1975–1995* (Houghton Mifflin, 1997), p. 15.

10. Seeing the Geometry in the Numbers

1. One reads in that preface to the Copernican treatise the somewhat surprising disclaimer that the theory to be set forth is "only a hypothesis." The disclaimer was written, we may assume, to reduce ecclesiastical interest in the work. A good try.
2. "Quel est celui de nous qui n'a pas dans ses jours d'ambition, révé le miracle d'une prose poétique, musicale sans rhythme et sans rime, assez souple et assez heurtée pour s'adapter aux mouvements lyrique de l'âme, aux ondulations de la rêverie, aux soubresauts de la conscience?" From the introduction to *Le Speen de Paris*, in Charles Baudelaire, *Oeuvres Complète* (Gallimard, 1958), p. 281.
3. From Virginia Woolf's essay "Impassioned Prose," initially published in *The Times Literary Supplement*, Sept. 16, 1926. It is reprinted in *Granite & Rainbow* (Harvest, 1958), p. 35.
4. Here are the four products:

$$3 \times 5 = 15$$
$$3 \times 6\sqrt{-1} = 18\sqrt{-1}$$
$$4\sqrt{-1} \times 5 = 20\sqrt{-1}$$
$$4\sqrt{-1} \times 6\sqrt{-1} = -24.$$

So,

$$(3 + 4\sqrt{-1}) \times (5 + 6\sqrt{-1}) = 15 + 18\sqrt{-1} + 20\sqrt{-1} - 24 = -9 + 38\sqrt{-1}.$$

5. Isaac Newton, *Universal Arithmetick*, p. 2.
6. This sentence encapsulates all the essential mathematical content of the high school subject of trigonometry. Of course, that subject also has the Adam-like pleasure of naming things (*sine*,

cosine, *tangent*, etc.) and the handy appurtenance of "trig tables" (these archaic devices having been replaced by calculators with trig functions).

11. The Literature of Discovery of Geometry in Numbers

1. We find hints, for example, in the writings of Viète, and more explicitly in Albert Girard's 1629 treatise *Invention nouvelle en l'algèbre*, which gives the trigonometric solutions of cubic equations.
2. This is the *doubling formula* in trigonometry. The coordinates (x,y) of the point C are the cosine and sine, respectively, of the double of the angle α in the diagram, and the doubling formula says that if c and s are the cosine and sine, respectively, of an angle α, then $c^2 - s^2$ and $2cs$ are the cosine and sine, respectively, of the double of α.
3. More precisely, this is how the analogy appears in the English version of his article, as published in *The Philosophical Transactions of the Royal Society of London from their Commencement in 1665 to the Year 1800; Abridged*, vol. 8 (1735–43), ed. Charles Hutton, George Shaw, and Richard Pearson (London, 1809). For readers who wish to make the comparison between De Moivre's text and my discussion of it, note that I have changed his letters a and b to c and s (to remind us that they are the cosine and sine of something), fixed the radius r to be 1, and transformed his expressions somewhat, using the "trig formula" $c^2 + s^2 = 1$.
4. De Moivre, "Reduction of Radicals." The numerical example $81 + \sqrt{-2700}$ is a curious choice to have made if one wanted to demonstrate the necessity of a "table of sines."
5. The quotation is from the English translation of an extract of Wallis's Latin text cited in D. E. Smith, *Source Book in Mathematics*, p. 48.
6. André Weil, "De la métaphysique aux mathématiques," in *André Weil: Oeuvres Scientifiques Collected Papers*, vol. 2 (1951–64) (Springer, 1979), pp. 408–12. A version of the quotation given in the text can be found in a letter André Weil

wrote to his sister, Simone Weil, some two decades earlier. See "Une lettre et un extrait de lettre à Simone Weil," ibid., vol. 1, pp. 244–55.

7. For an assessment of the validity of this and other published biographical information on De Moivre, see Helen Walker, "Abraham De Moivre," in *Scripta Mathematica*, vol. 2 (1933–34) (Yeshiva College, 1934).
8. L. Euler, "Recherches sur les racines imaginaires des équations," *Histoire de l'Academie Royale des Sciences et Belles Lettres* (Berlin), vol. 5, 1749, pp. 222–88. A translation into English of an extract of this is in Smith, pp. 452–54.
9. For a detailed discussion of this literature and its history, see chap. 3 in Paul Nahin, *The Story of* $\sqrt{-1}$ (Princeton Univ. Press, 1998). See also the review by B. Blank in *Notices of the American Mathematical Society* 46 (Nov. 1999): 1233–36.
10. Gauss would go on to publish four different proofs of this in the course of his lifetime.
11. J.-F. Français's article appeared in *Annales de mathématiques pures et appliquées* 4, no. 2 (Sept. 1813), published in Nîmes. Hereafter I refer to this journal as *Annales*.
12. Argand's note has, but for punctuation, the same title as his 1806 treatise: "Essai sur une manière de représenter les quantités imaginaires, dans les constructions géométriques," *Annales* 4, no. 5 (Nov. 1813). For an English translation of an extract of Wessel's paper, see D. E. Smith, *A Source Book in Mathematics*, pp. 55–66.
13. This is the theorem proved by Gauss in his thesis, discussed at the beginning of this section.
14. Gergonne had *many* friends; see the memoir by Niels Nielsen, *Géométres Français sous la Révolution* (Levin & Munksgaard, 1929).
15. *Annales*, p. 70.
16. Ibid., p. 229.
17. "M. Servois compterait-il donc pour peu de voir enfin l'analise algébrique débarassée de ces formes inintelligibles et mysterieuses, de ces *non sense* qui la déparent et en font, pour ainsi dire, une sort de science cabalistique?"
18. In Wessel's as well as Français's article one finds the rudiments

of polar coordinates. Français talks of line elements in the plane of a given length (what we call the *magnitude*) r, making a specific angle with the horizontal (what we call the *phase*) α. He never seems to say explicitly that he wants these directed line elements to have their beginning point at the origin, but one can only assume that he means to do so. When he "identifies" the line elements with complex numbers, he notes that to multiply two such, viewed as complex numbers, is the same as to multiply their rs and add their αs.

19. "*Corollaire 3.* Les quantités dite *imaginaires* sont donc tout aussi réelle que les quantités positives et les quantités negatives, et n'en different que par leur position, qui est perpendiculaire à celle de ces dernières."

12. Understanding Algebra via Geometry

1. Imagine the mix-up that occurs when the minister confuses twins at a baptism, as in Grace Dane Mazur's novel *Trespass* (Graywolf, 2002).
2. Here are brief hints for the proof: Let z denote a cube root of $c/2 + \sqrt{c^2/4 - b^3/27}$ and z' denote its conjugate; then (noting that b is positive), $zz' = b/3$. Now evaluate $(z + z')^3$.
3. Quoted from Carl Boyer, *History of Mathematics*, rev. Uta Merzbach (Wiley, 1989), p. 310. See Boyer's account for further discussion of the history of trigonometric solution of equations.
4. We are dealing with the case where $b = 3$ and $c = 1$. So Dal Ferro's expression becomes the sum of the cube root of $D = (1 + \sqrt{-3})/2$ and its conjugate $\bar{D} = (1 - \sqrt{-3})/2$. The complex number $(1 + \sqrt{-3})/2$ (expressed in polar coordinates) has magnitude 1 and phase 60 degrees. So its three cube roots all have magnitude 1 and have phases 20, 140, and 260 degrees. For the first of these cube roots, the first displayed triangle in the text diagram is relevant: the sum of this cube root and its conjugate is then twice the length of the base of that triangle; namely 1.87939. . . . For the second and third cube roots, the sum of cube root and corresponding conjugate for each corresponding triangle is twice the length of its base, with appropri-

ate sign: −1.532088 . . . and −0.347296 . . . , respectively. So the three solutions of the equation $X^3 = 3X + 1$ are given by these three numbers. Of course, if you want the solutions to greater accuracy (i.e., more digits in their decimal expansion) you need to compute (twice) the lengths of the bases of the three triangles in the diagram to greater accuracy. Here they are to twelve decimal places:

$$X = 1.879385241572\ldots$$
$$X = -1.532088886238\ldots$$
$$X = -0.347296355334\ldots$$

Check that these values of X do the trick!

5. For an excellent discussion of this aspect of poetry, see Robert Pinsky, *The Sounds of Poetry* (Farrar, Straus and Giroux, 1998).
6. Helen Vendler, *The Art of Shakespeare's Sonnets* (Harvard Univ. Press, 1997) p. 3.
7. This is the opening of Emily Dickinson's poem 372, in *The Poems of Emily Dickinson*, ed. R. W. Franklin, vol. 2 (Harvard Univ. Press, 1998).
8. Immanuel Kant, "Critique of Teleological Judgment," in *The Critique of Judgment*, trans. Werner Pluhar (Hackett, 1987), p. 244, para. 63, n. 23. Regarding the quotation "Mathematics is pure poetry," from Kant's *Opus Postumum*, see David Lachterman's use of this phrase in *Ethics of Geometry*, p. xiv. A recent book on Kant's aesthetics of mathematics is M. Aissen-Crewett, *Mathematik ist reine Dichtung* (Potsdam Univ. Press, 2000).
9. From Oscar Wilde's preface to his *The Picture of Dorian Gray* (Oxford Univ. Press, 1981), p. xxiii.
10. "Whatever It Is, Wherever You Are," p. 63.
11. From *The Collected Poems of Wallace Stevens* (Knopf, 1978). I was reminded of this tripartite poem by reading Ellen Bryant Voigt's inspiring collection of essays on the sounds and the sense of lyric poetry, *The Flexible Lyric* (Univ. of Georgia Press, 1999).

BIBLIOGRAPHY

Accounts of Imaginary Numbers, Accessible to Readers with No Background in Mathematics

Number: The Language of Science, by Tobias Dantzig (Free Press, 1954), is the great classic. Beginning with a discussion of the "number sense" shared by all of us, Dantzig's wonderful book develops the idea of number systems and gets to complex numbers (in chapter 10, "The Domain of Number").

Mathematics for the Millions, by Lancelot Hogben (W. W. Norton, 1937), is a book "for the million or so who have given up hope of learning [math] through the usual channels." Hogben's book has a broader mission than Dantzig's but does a fine job in its treatment of complex numbers (in chapter 7, "The Dawn of Nothing, or How Algebra Began"). His very useful "Instructions for Readers" tells one how to read a math book.

One, Two, Three, . . . Infinity, by George Gamow (Bantam, 1967), has a marvelous seven-page account of complex numbers in the chapter "The Mysterious $\sqrt{-1}$." In it Gamow describes a treasure-hunt problem that can be elegantly solved using multiplication of simple complex numbers and, in turn, demonstrates the power of the complex plane.

QED: The Strange Theory of Light and Matter, by Richard P. Feynman (Princeton Univ. Press, 1985), is ostensibly a book about physics, but it has a vivid explanation of complex numbers, central to Feynman's discussion of quantum electrodynamics. Start, for example, by looking at his half-page footnote (p. 63) about numbers.

Where Mathematics Comes From, by George Lakoff and Rafael Núñez (Basic Books, 2000), has a fine discussion of complex numbers and the imaginative work it takes for us to be at home with them. See specifically their "Case Study 3: What Is *i*?" in part VI.

Histories of Various Aspects of Mathematics, Accessible to Readers with a High School Background in Math

Berlinghoff, W. P., and F. Q. Gouvêa, *Math Through the Ages* (Oxton House, 2002).

Boyer, C., rev. Uta C. Merzbach, *History of Mathematics* (Wiley, 1989).

Grattan-Guinness, I., *The Norton History of the Mathematical Sciences* (Norton, 1997).

Kaplan, B., *The Nothing That Is: A Natural History of Zero* (Oxford Univ. Press, 2000).

Klein, J., *Greek Mathematical Thought and the Origin of Algebra* (MIT Press, 1968).

Neugebauer, O., *The Exact Sciences in Antiquity* (Brown Univ. Press, 1957).

Histories of Various Aspects of Mathematics Requiring More Mathematical Background

Cajori, F., *A History of Mathematical Notations*, vols. 1, 2 (Open Court, 1929).

Fowler, D., *The Mathematics of Plato's Academy* (Oxford Univ. Press, 1987).

Ore, O., *Number Theory and Its History* (McGraw-Hill, 1948).

Nahin, P., *An Imaginary Tale: The Story of* $\sqrt{-1}$ (Princeton Univ. Press, 1998).

Weil, A., *Number Theory: An Approach Through History from Hammurapi to Legendre* (Birkhäuser, 1983).

ACKNOWLEDGMENTS

What a pleasure it is to thank friends for their conversation and guidance. Some of the conversations connected to the subject of my book began decades ago—with Eva Brann, Persi Diaconis, Bob Kaplan, Ellen Kaplan, David Kazhdan, Valentin Poenaru, Gabe Stolzenberg, and Gretchen, Zeke, and Joe Mazur—and are still going on, as is my conversation with Licsi Szatmari, a newer member of my family. Conversations with Michel Chaouli and Elaine Scarry got me started writing this book.

I am grateful that people so generously took the time to go through early drafts, helping me with their corrections, their insights, their scholarship, and their advice. For all this and more, I am delighted to thank Manjul Bhargava, Stephen Booth, Eva Brann, Carl Brownsberger, Michel Chaouli, Capi Corrales, David Cox, Joseph Daubin, Persi Diaconis, Zvi Dor-Ner, Fernando Gouvêa, Michael Harris, Bob Kaplan, David Kazhdan, George Lakoff, Gretchen Mazur, Curt McMullen, Peter Pesic, Simon Singh, Paul Solman, Dick Teresi, and Barbara Tversky.

I was also most fortunate to have had suggestions, and crucial help, from Avner Ash, Don Fanger, Sorana Froda, David Gewanter, Victor Guillemin, Susan Holmes, Michelle Jaffe, Sarah Kafatou, Elena Mantovan, Federica La Nave, and Paula Panich. Thank you.

I thank Eric Simonoff and Jonathan Galassi for believing that this actually could be a book, and for their valuable counsel; I appreciate the immense effort that James Wilson and the rest of the staff at FSG spent on production. I am indebted to Linda Strange for the copyediting, Mary Reilly for the diagrams, Irene Minder for general technical help (and much more), and Anne Farma for liberating me from the tangles of my computer.

INDEX

PERMISSIONS ACKNOWLEDGMENTS

Grateful acknowledgment is made for permission to reprint the following:

pp. 43–44: "Adam's Curse" is reprinted with the permission of Scribner, an imprint of Simon and Schuster Adult Publishing Group, from *The Collected Works of W. B. Yeats, Volume 1: The Poems, Revised,* edited by Richard J. Finneran (New York: Scribner, 1997).

p. 56: "There or Then" from *Figurehead* by John Hollander, copyright © 1999 by John Hollander. Used by permission of Alfred A. Knopf, a division of Random House, Inc.

p. 76: "Elegy" from *Second Four Books of Poems* by W. S. Merwin, copyright © 1970 by W. S. Merwin, reprinted by permission of The Wylie Agency, Inc.

p. 89: "Unmarked Boxes" from *The Essential Rumi*, tr. Coleman Barks, with John Moyne (San Francisco: HarperSanFrancisco, 1996). Copyright © 1996 Coleman Barks.